D0241538

2016 SQA Past Papers With Answers

National 5
PHYSICS

National 5 PHYSICS

2014, 2015 & 2016 Exams

HODDER
GIBSON
AN HACHETTE UK COMPANY

This book contains the official SQA 2014, 2015 and 2016 Exams for National 5 Physics, with associated SQA-approved answers modified from the official marking instructions that accompany the paper.

In addition the book contains study skills advice. This advice has been specially commissioned by Hodder Gibson, and has been written by experienced senior teachers and examiners in line with the new National 5 syllabus. This is not SQA material but has been devised to provide further guidance for National 5 examinations.

Hodder Gibson is grateful to the copyright holders, as credited on the final page of the Answer Section, for permission to use their material. Every effort has been made to trace the copyright holders and to obtain their permission for the use of copyright material. Hodder Gibson will be happy to receive information allowing us to rectify any error or omission in future editions.

Hachette UK's policy is to use papers that are natural, renewable and recyclable products and made from wood grown in sustainable forests. The logging and manufacturing processes are expected to conform to the environmental regulations of the country of origin.

Orders: please contact Bookpoint Ltd, 130 Park Drive, Milton Park, Abingdon, Oxon OX14 4SE. Telephone: (44) 01235 827720. Fax: (44) 01235 400454. Lines are open 9.00–5.00, Monday to Saturday, with a 24-hour message answering service. Visit our website at www.hoddereducation.co.uk. Hodder Gibson can be contacted direct on: Tel: 0141 333 4650; Fax: 0141 404 8188; email: hoddergibson@hodder.co.uk

This collection first published in 2016 by
Hodder Gibson, an imprint of Hodder Education,
An Hachette UK Company
211 St Vincent Street
Glasgow G2 5QY

National 5 2014, 2015 and 2016 Exam Papers and Answers © Scottish Qualifications Authority. Study Skills section © Hodder Gibson. All rights reserved. Apart from any use permitted under UK copyright law, no part of this publication may be reproduced or transmitted in any form or by any means, electronic or mechanical, including photocopying and recording, or held within any information storage and retrieval system, without permission in writing from the publisher or under licence from the Copyright Licensing Agency Limited. Further details of such licences (for reprographic reproduction) may be obtained from the Copyright Licensing Agency Limited, Saffron House, 6–10 Kirby Street, London EC1N 8TS.

Typeset by Aptara, Inc.

Printed in the UK

A catalogue record for this title is available from the British Library

ISBN: 978-1-4718-9119-9

3 2 1

2017 2016

Introduction

Study Skills – what you need to know to pass exams!

Pause for thought

Many students might skip quickly through a page like this. After all, we all know how to revise. Do you really though?

Think about this:

"IF YOU ALWAYS DO WHAT YOU ALWAYS DO, YOU WILL ALWAYS GET WHAT YOU HAVE ALWAYS GOT."

Do you like the grades you get? Do you want to do better? If you get full marks in your assessment, then that's great! Change nothing! This section is just to help you get that little bit better than you already are.

There are two main parts to the advice on offer here. The first part highlights fairly obvious things but which are also very important. The second part makes suggestions about revision that you might not have thought about but which WILL help you.

Part 1

DOH! It's so obvious but …

Start revising in good time

Don't leave it until the last minute – this will make you panic.

Make a revision timetable that sets out work time AND play time.

Sleep and eat!

Obvious really, and very helpful. Avoid arguments or stressful things too – even games that wind you up. You need to be fit, awake and focused!

Know your place!

Make sure you know exactly **WHEN and WHERE** your exams are.

Know your enemy!

Make sure you know what to expect in the exam.

How is the paper structured?

How much time is there for each question?

What types of question are involved?

Which topics seem to come up time and time again?

Which topics are your strongest and which are your weakest?

Are all topics compulsory or are there choices?

Learn by DOING!

There is no substitute for past papers and practice papers – they are simply essential! Tackling this collection of papers and answers is exactly the right thing to be doing as your exams approach.

Part 2

People learn in different ways. Some like low light, some bright. Some like early morning, some like evening / night. Some prefer warm, some prefer cold. But everyone uses their BRAIN and the brain works when it is active. Passive learning – sitting gazing at notes – is the most INEFFICIENT way to learn anything. Below you will find tips and ideas for making your revision more effective and maybe even more enjoyable. What follows gets your brain active, and active learning works!

Activity 1 – Stop and review

Step 1

When you have done no more than 5 minutes of revision reading STOP!

Step 2

Write a heading in your own words which sums up the topic you have been revising.

Step 3

Write a summary of what you have revised in no more than two sentences. Don't fool yourself by saying, "I know it, but I cannot put it into words". That just means you don't know it well enough. If you cannot write your summary, revise that section again, knowing that you must write a summary at the end of it. Many of you will have notebooks full of blue/black ink writing. Many of the pages will not be especially attractive or memorable so try to liven them up a bit with colour as you are reviewing and rewriting. **This is a great memory aid, and memory is the most important thing.**

Activity 2 – Use technology!

Why should everything be written down? Have you thought about "mental" maps, diagrams, cartoons and colour to help you learn? And rather than write down notes, why not record your revision material?

What about having a text message revision session with friends? Keep in touch with them to find out how and what they are revising and share ideas and questions.

Why not make a video diary where you tell the camera what you are doing, what you think you have learned and what you still have to do? No one has to see or hear it, but the process of having to organise your thoughts in a formal way to explain something is a very important learning practice.

Be sure to make use of electronic files. You could begin to summarise your class notes. Your typing might be slow, but it will get faster and the typed notes will be easier to read than the scribbles in your class notes. Try to add different fonts and colours to make your work stand out. You can easily Google relevant pictures, cartoons and diagrams which you can copy and paste to make your work more attractive and **MEMORABLE**.

Activity 3 – This is it. Do this and you will know lots!

Step 1

In this task you must be very honest with yourself! Find the SQA syllabus for your subject (www.sqa.org.uk). Look at how it is broken down into main topics called MANDATORY knowledge. That means stuff you MUST know.

Step 2

BEFORE you do ANY revision on this topic, write a list of everything that you already know about the subject. It might be quite a long list but you only need to write it once. It shows you all the information that is already in your long-term memory so you know what parts you do not need to revise!

Step 3

Pick a chapter or section from your book or revision notes. Choose a fairly large section or a whole chapter to get the most out of this activity.

With a buddy, use Skype, Facetime, Twitter or any other communication you have, to play the game "If this is the answer, what is the question?". For example, if you are revising Geography and the answer you provide is "meander", your buddy would have to make up a question like "What is the word that describes a feature of a river where it flows slowly and bends often from side to side?".

Make up 10 "answers" based on the content of the chapter or section you are using. Give this to your buddy to solve while you solve theirs.

Step 4

Construct a wordsearch of at least 10 × 10 squares. You can make it as big as you like but keep it realistic. Work together with a group of friends. Many apps allow you to make wordsearch puzzles online. The words and phrases can go in any direction and phrases can be split. Your puzzle must only contain facts linked to the topic you are revising. Your task is to find 10 bits of information to hide in your puzzle, but you must not repeat information that you used in Step 3. DO NOT show where the words are. Fill up empty squares with random letters. Remember to keep a note of where your answers are hidden but do not show your friends. When you have a complete puzzle, exchange it with a friend to solve each other's puzzle.

Step 5

Now make up 10 questions (not "answers" this time) based on the same chapter used in the previous two tasks. Again, you must find NEW information that you have not yet used. Now it's getting hard to find that new information! Again, give your questions to a friend to answer.

Step 6

As you have been doing the puzzles, your brain has been actively searching for new information. Now write a NEW LIST that contains only the new information you have discovered when doing the puzzles. Your new list is the one to look at repeatedly for short bursts over the next few days. Try to remember more and more of it without looking at it. After a few days, you should be able to add words from your second list to your first list as you increase the information in your long-term memory.

FINALLY! Be inspired...

Make a list of different revision ideas and beside each one write **THINGS I HAVE** tried, **THINGS I WILL** try and **THINGS I MIGHT** try. Don't be scared of trying something new.

And remember – "FAIL TO PREPARE AND PREPARE TO FAIL!"

National 5 Physics

The exam

Duration: **2 hours**
Total marks: **110**

20 marks are awarded for 20 **multiple-choice questions** – completed on an answer grid.

90 marks are awarded for **written answers** – completed in the space provided after each question or on graph paper.

Approximately one third of the 110 marks are allocated to questions from each unit.

The National 5 Physics course consists of **three units**:

- Unit 1 – Electricity and Energy
- Unit 2 – Waves and Radiation
- Unit 3 – Dynamics and Space

General exam advice

There are 110 marks in total, and you have two hours to complete the paper. This works out at just over one minute per mark, so a 10 mark question would take roughly 11 minutes.

Be aware of how much time you spend on each question. For example, DO NOT spend 10 minutes on a question worth only three marks, especially when you haven't completed the rest of the questions – you can always return to the question later if there's time.

The best method for getting used to National 5 exam questions is to attempt as many exam type questions as possible, **and check your answers**. If you find a wrong answer, **find out why it is wrong** and then try similar questions until you can answer them correctly.

Specific exam advice

Advice for answering multiple-choice questions (Section 1) (20 marks)

Each question has five possible choices of answers. **Only one answer is correct.**

Multiple-choice questions are designed to test a range of skills, e.g.

- knowledge and understanding of the course
- using equations
- selecting correct statements from a list
- selecting and analysing information from a diagram.

It is important to **practise** as many **multiple-choice questions** as possible, to get used to the "style" and types of questions.

Do not try to work out all of the answers to multiple-choice questions in your head. Instead, when the question is complicated, write down notes and work on scrap paper (provided by the invigilator) or use the blank pages at the end of the question paper.

Do not use the answer grid for working, and remember to cross out your multiple-choice rough working when you have finished.

You can also make notes beside the actual question, if this helps, but **not** on the answer grid.

Advice for answering written questions (Section 2) (90 marks)

These questions test several different skills.

The majority of these marks test your **knowledge and understanding** of the course.

There are also questions which test different skills, like selecting information, analysing information, predicting results, and commenting on experimental results.

There are usually around 12–14 questions in Section 2. There are different types of questions, which include:

- Questions testing your **knowledge of the course**, sometimes applied to particular applications. More than half of the 90 marks in Section 2 are for this type of question.

- Questions (usually a maximum of two) involving **physics content not in the course** but explained in the question, usually including an equation which you are asked to use with data.

- A question testing your **scientific reading skills**, where you will be asked about a scientific report or passage. The question might include a calculation.

- **"Open-ended" questions** (a maximum of 2 per exam, three marks each), which usually discuss a physics phenomenon and ask you to explain it using your knowledge of physics. You have to think about the issue and try to give a step-by-step answer – there may be more than one area of physics used to answer this type of question. These questions allow you to use your knowledge and problem-solving skills. Be careful not to spend longer than necessary on these three mark questons.

- Questions testing practical skills usually based on **practical or experimental work**, which may have tables of results or graphs (or both) which have to be used to obtain information needed to answer the question. You could be asked to identify a problem with the results, or to suggest an improvement to the experiment.

Things to remember when answering questions

Using equations

More than half of the total marks awarded in Section 2 are for being able to calculate answers using an equation (relationship) from the **"Relationship Sheet"** which is supplied with the exam paper.

These questions are usually worth three marks. To obtain the full three marks for these questions, your final answer must be correct.

There are three separate marks awarded for the stages of the working:

- Write down the correct equation needed to calculate the answer from the Relationship Sheet – **1 mark**.
- Show that the correct values are substituted into the equation – **1 mark**.
- Show the final answer, including the correct unit – **1 mark**.

If the unit is wrong or missing, you will lose the final mark!

Other important areas to remember and practise are:

Units

The units of measurement in the National 5 Physics course are based on the International System of Units. Make sure that you use the correct unit following a calculation in your final answer.

Prefixes

A prefix produces a multiple of the unit in powers of ten, e.g. 10^{-6} is 0·000001. It is named "micro" and has the symbol "μ". Make sure to practise and get used to all prefixes.

Scientific notation

This is used in the exam to write very large or very small numbers, to avoid writing or using strings of numbers in an answer or calculation.

You need to be familiar with how to enter and use numbers in scientific notation on **your** calculator – make sure that you have used your calculator often before the exam to get used to it.

Significant figures

When calculating a value using an equation, take care not to give too many significant figures in the final answer. If there are intermediate steps in a calculation, you can keep numbers in your calculator which have too many significant figures. You should always round your answer to give no more than the smallest number of significant figures which appear in the data given in the question.

$$\text{E.g. } \frac{42\cdot74}{2\cdot59} = 16\cdot5019305$$

If the smallest number of significant figures relating to the data used from the question was three, then round this answer to 16·5.

Examples:

- 20 has 1 significant figure
- 40·0 has 3 significant figures
- 0·000604 has 3 significant figures
- $4\cdot30 \times 10^4$ has 3 significant figures
- 6200 has 2 significant figures

Good luck!

Remember that the rewards for passing National 5 Physics are well worth it! Your pass will help you get the future you want for yourself. In the exam, be confident in your own ability. If you are not sure how to answer a question, trust your instincts and just give it a go anyway. Keep calm and don't panic! GOOD LUCK!

NATIONAL 5

2014

National Qualifications 2014

X757/75/02

**Physics
Section 1—Questions**

THURSDAY, 22 MAY
9:00 AM – 11:00 AM

Instructions for the completion of Section 1 are given on *Page two* of your question and answer booklet X757/75/01.

Record your answers on the answer grid on *Page three* of your question and answer booklet.

Reference may be made to the Data Sheet on *Page two* of this booklet and to the Relationship Sheet X757/75/11.

Before leaving the examination room you must give your question and answer booklet to the Invigilator; if you do not, you may lose all the marks for this paper.

DATA SHEET

Speed of light in materials

Material	Speed in $m\,s^{-1}$
Air	$3{\cdot}0 \times 10^8$
Carbon dioxide	$3{\cdot}0 \times 10^8$
Diamond	$1{\cdot}2 \times 10^8$
Glass	$2{\cdot}0 \times 10^8$
Glycerol	$2{\cdot}1 \times 10^8$
Water	$2{\cdot}3 \times 10^8$

Gravitational field strengths

	Gravitational field strength on the surface in $N\,kg^{-1}$
Earth	9·8
Jupiter	23
Mars	3·7
Mercury	3·7
Moon	1·6
Neptune	11
Saturn	9·0
Sun	270
Uranus	8·7
Venus	8·9

Specific latent heat of fusion of materials

Material	Specific latent heat of fusion in $J\,kg^{-1}$
Alcohol	$0{\cdot}99 \times 10^5$
Aluminium	$3{\cdot}95 \times 10^5$
Carbon dioxide	$1{\cdot}80 \times 10^5$
Copper	$2{\cdot}05 \times 10^5$
Iron	$2{\cdot}67 \times 10^5$
Lead	$0{\cdot}25 \times 10^5$
Water	$3{\cdot}34 \times 10^5$

Specific latent heat of vaporisation of materials

Material	Specific latent heat of vaporisation in $J\,kg^{-1}$
Alcohol	$11{\cdot}2 \times 10^5$
Carbon dioxide	$3{\cdot}77 \times 10^5$
Glycerol	$8{\cdot}30 \times 10^5$
Turpentine	$2{\cdot}90 \times 10^5$
Water	$22{\cdot}6 \times 10^5$

Speed of sound in materials

Material	Speed in $m\,s^{-1}$
Aluminium	5200
Air	340
Bone	4100
Carbon dioxide	270
Glycerol	1900
Muscle	1600
Steel	5200
Tissue	1500
Water	1500

Specific heat capacity of materials

Material	Specific heat capacity in $J\,kg^{-1}\,{}^{\circ}C^{-1}$
Alcohol	2350
Aluminium	902
Copper	386
Glass	500
Ice	2100
Iron	480
Lead	128
Oil	2130
Water	4180

Melting and boiling points of materials

Material	Melting point in °C	Boiling point in °C
Alcohol	−98	65
Aluminium	660	2470
Copper	1077	2567
Glycerol	18	290
Lead	328	1737
Iron	1537	2737

Radiation weighting factors

Type of radiation	Radiation weighting factor
alpha	20
beta	1
fast neutrons	10
gamma	1
slow neutrons	3
X-rays	1

SECTION 1

1. The voltage of an electrical supply is a measure of the

    A   resistance of the circuit

    B   speed of the charges in the circuit

    C   power developed in the circuit

    D   energy given to the charges in the circuit

    E   current in the circuit.

2. Four circuit symbols, W, X, Y and Z, are shown.

    Which row identifies the components represented by these symbols?

	W	X	Y	Z
A	battery	ammeter	resistor	variable resistor
B	battery	ammeter	fuse	resistor
C	lamp	ammeter	variable resistor	resistor
D	lamp	voltmeter	resistor	fuse
E	lamp	voltmeter	variable resistor	fuse

[Turn over

3. A student suspects that ammeter A_1 may be inaccurate. Ammeter A_2 is known to be accurate.

Which of the following circuits should be used to compare the reading on A_1 with A_2?

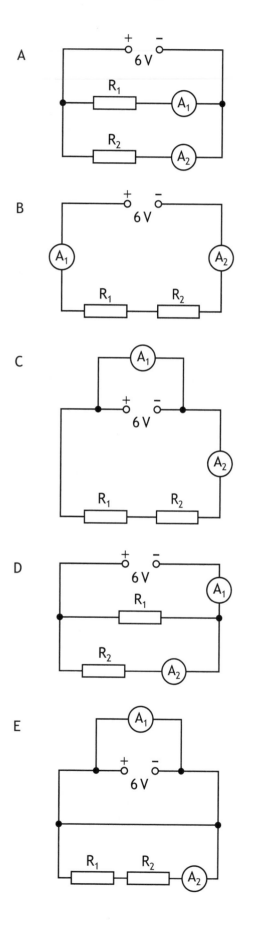

4. A ball of mass 0·50 kg is released from a height of 1·00 m and falls towards the floor.

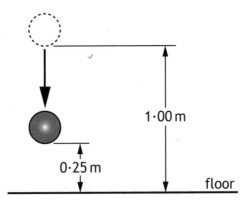

Which row in the table shows the gravitational potential energy and the kinetic energy of the ball when it is at a height of 0·25 m from the floor?

	Gravitational potential energy (J)	Kinetic energy (J)
A	0·12	0·12
B	1·2	1·2
C	1·2	3·7
D	3·7	1·2
E	4·9	1·2

5. The pressure of a fixed mass of gas is $6·0 \times 10^5$ Pa.

The temperature of the gas is 27 °C and the volume of the gas is 2·5 m^3.

The temperature of the gas increases to 54 °C and the volume of the gas increases to 5·0 m^3.

What is the new pressure of the gas?

A $2·8 \times 10^5$ Pa

B $3·3 \times 10^5$ Pa

C $6·0 \times 10^5$ Pa

D $1·1 \times 10^6$ Pa

E $1·3 \times 10^6$ Pa

[Turn over

6. A student is investigating the relationship between the volume and the kelvin temperature of a fixed mass of gas at constant pressure.

Which graph shows this relationship?

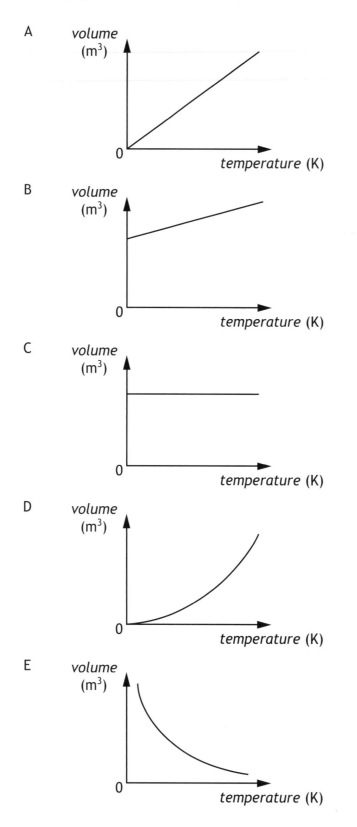

7. A liquid is heated from 17 °C to 50 °C. The temperature rise in kelvin is

   A   33 K

   B   67 K

   C   306 K

   D   340 K

   E   579 K.

8. The period of vibration of a guitar string is 8 ms.

   The frequency of the sound produced by the guitar string is

   A   0·125 Hz

   B   12·5 Hz

   C   125 Hz

   D   800 Hz

   E   8000 Hz.

9. A student makes the following statements about microwaves and radio waves.

   I   In air, microwaves travel faster than radio waves.

   II   In air, microwaves have a longer wavelength than radio waves.

   III   Microwaves and radio waves are both members of the electromagnetic spectrum.

   Which of these statements is/are correct?

   A   I only

   B   III only

   C   I and II only

   D   I and III only

   E   II and III only

10. Which row describes alpha (α), beta (β) and gamma (γ) radiations?

	α	β	γ
A	helium nucleus	electromagnetic radiation	electron from the nucleus
B	helium nucleus	electron from the nucleus	electromagnetic radiation
C	electron from the nucleus	helium nucleus	electromagnetic radiation
D	electromagnetic radiation	helium nucleus	electron from the nucleus
E	electromagnetic radiation	electron from the nucleus	helium nucleus

11. A sample of tissue is irradiated using a radioactive source.

A student makes the following statements about the sample.

  I   The equivalent dose received by the sample is reduced by shielding the sample with a lead screen.

  II The equivalent dose received by the sample is increased as the distance from the source to the sample is increased.

  III The equivalent dose received by the sample is increased by increasing the time of exposure of the sample to the radiation.

Which of these statements is/are correct?

A   I only

B   II only

C   I and II only

D   II and III only

E   I and III only

12. The half-life of a radioactive source is 64 years.

In 2 hours, $1 \cdot 44 \times 10^8$ radioactive nuclei in the source decay.

What is the activity of the source in Bq?

A 2×10^4

B 4×10^4

C $1 \cdot 2 \times 10^6$

D $2 \cdot 25 \times 10^6$

E $7 \cdot 2 \times 10^7$

13. A student makes the following statements about the fission process in a nuclear power station.

  I   Electrons are used to bombard a uranium nucleus.

  II Heat is produced.

  III The neutrons released can cause other nuclei to undergo fission.

Which of these statements is/are correct?

A   I only

B   II only

C   III only

D   I and II only

E   II and III only

14. Which of the following contains two vectors and one scalar quantity?

    A   Acceleration, mass, displacement

    B   Displacement, force, velocity

    C   Time, distance, force

    D   Displacement, velocity, acceleration

    E   Speed, velocity, distance

15. A vehicle follows a course from R to T as shown.

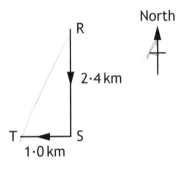

The total journey takes 1 hour.

Which row in the table gives the average speed and the average velocity of the vehicle for the whole journey?

	Average speed	Average velocity
A	2·6 km h^{-1} (023)	3·4 km h^{-1}
B	2·6 km h^{-1}	3·4 km h^{-1} (203)
C	3·4 km h^{-1} (203)	2·6 km h^{-1}
D	3·4 km h^{-1}	2·6 km h^{-1} (023)
E	3·4 km h^{-1}	2·6 km h^{-1} (203)

16. A force of 10 N acts on an object for 2 s.

During this time the object moves a distance of 3 m.

The work done on the object is

    A   6·7 J

    B   15 J

    C   20 J

    D   30 J

    E   60 J.

17. Catapults are used by anglers to project fish bait into water.

A technician designs a catapult for this use.

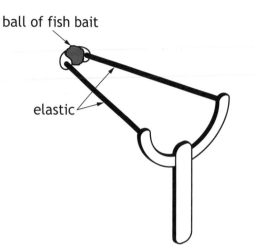

ball of fish bait

elastic

Pieces of elastic of different thickness are used to provide a force on the ball.

Each piece of elastic is the same length.

The amount of stretch given to each elastic is the same each time.

The force exerted on the ball increases as the thickness of the elastic increases.

Which row in the table shows the combination of the thickness of elastic and mass of ball that produces the greatest acceleration?

	Thickness of elastic (mm)	Mass of ball (kg)
A	5	0·01
B	10	0·01
C	10	0·02
D	15	0·01
E	15	0·02

18. A spacecraft completes the last stage of its journey back to Earth by parachute, falling with constant speed into the sea.

    The spacecraft falls with constant speed because

    A   the gravitational field strength of the Earth is constant near the Earth's surface

    B   it has come from space where the gravitational field strength is almost zero

    C   the air resistance is greater than the weight of the spacecraft

    D   the weight of the spacecraft is greater than the air resistance

    E   the air resistance is equal to the weight of the spacecraft.

19. A ball is released from point **Q** on a curved rail, leaves the rail horizontally at R and lands 1 s later.

    The ball is now released from point **P**.

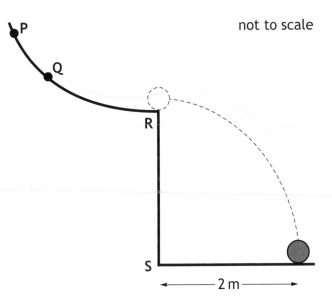

not to scale

    Which row describes the motion of the ball after leaving the rail?

	Time to land after leaving rail	Distance from S to landing point
A	1 s	less than 2 m
B	less than 1 s	more than 2 m
C	1 s	more than 2 m
D	less than 1 s	2 m
E	more than 1 s	more than 2 m

20. A solid substance is placed in an insulated flask and heated continuously with an immersion heater.

The graph shows how the temperature of the substance in the flask changes in time.

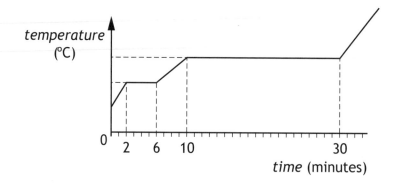

After 5 minutes the substance is a

A   solid

B   liquid

C   gas

D   mixture of solid and liquid

E   mixture of liquid and gas.

[END OF SECTION 1. NOW ATTEMPT THE QUESTIONS IN SECTION 2 OF YOUR QUESTION AND ANSWER BOOKLET]

National Qualifications 2014

X757/75/11

**Physics
Relationships Sheet**

THURSDAY, 22 MAY

9:00 AM – 11:00 AM

$$E_p = mgh$$

$$E_k = \tfrac{1}{2}mv^2$$

$$Q = It$$

$$V = IR$$

$$R_T = R_1 + R_2 + \dots$$

$$\frac{1}{R_T} = \frac{1}{R_1} + \frac{1}{R_2} + \dots$$

$$V_2 = \left(\frac{R_2}{R_1 + R_2}\right)V_s$$

$$\frac{V_1}{V_2} = \frac{R_1}{R_2}$$

$$P = \frac{E}{t}$$

$$P = IV$$

$$P = I^2 R$$

$$P = \frac{V^2}{R}$$

$$E_h = cm\Delta T$$

$$p = \frac{F}{A}$$

$$\frac{pV}{T} = \text{constant}$$

$$p_1 V_1 = p_2 V_2$$

$$\frac{p_1}{T_1} = \frac{p_2}{T_2}$$

$$\frac{V_1}{T_1} = \frac{V_2}{T_2}$$

$$d = vt$$

$$v = f\lambda$$

$$T = \frac{1}{f}$$

$$A = \frac{N}{t}$$

$$D = \frac{E}{m}$$

$$H = Dw_R$$

$$\dot{H} = \frac{H}{t}$$

$$s = vt$$

$$d = \bar{v}\,t$$

$$s = \bar{v}\,t$$

$$a = \frac{v - u}{t}$$

$$W = mg$$

$$F = ma$$

$$E_w = Fd$$

$$E_h = ml$$

Additional Relationships

Circle

circumference $= 2\pi r$

area $= \pi r^2$

Sphere

area $= 4\pi r^2$

volume $= \frac{4}{3}\pi r^3$

Trigonometry

$\sin \theta = \dfrac{\text{opposite}}{\text{hypotenuse}}$

$\cos \theta = \dfrac{\text{adjacent}}{\text{hypotenuse}}$

$\tan \theta = \dfrac{\text{opposite}}{\text{adjacent}}$

$\sin^2 \theta + \cos^2 \theta = 1$

Electron Arrangements of Elements

Key

Atomic number
Symbol
Electron arrangement
Name

Group 1

Atomic number	Symbol	Name	Electron arrangement
1	H	Hydrogen	1
3	Li	Lithium	2,1
11	Na	Sodium	2,8,1
19	K	Potassium	2,8,8,1
37	Rb	Rubidium	2,8,18,8,1
55	Cs	Caesium	2,8,18,18,8,1
87	Fr	Francium	2,8,18,32,18,8,1

Group 2

Atomic number	Symbol	Name	Electron arrangement
4	Be	Beryllium	2,2
12	Mg	Magnesium	2,8,2
20	Ca	Calcium	2,8,8,2
38	Sr	Strontium	2,8,18,8,2
56	Ba	Barium	2,8,18,18,8,2
88	Ra	Radium	2,8,18,32,18,8,2

Transition Elements

Group	Atomic number	Symbol	Name	Electron arrangement
(3)	21	Sc	Scandium	2,8,9,2
(3)	39	Y	Yttrium	2,8,18,9,2
(3)	57	La	Lanthanum	2,8,18,18,9,2
(3)	89	Ac	Actinium	2,8,18,32,18,9,2
(4)	22	Ti	Titanium	2,8,10,2
(4)	40	Zr	Zirconium	2,8,18,10,2
(4)	72	Hf	Hafnium	2,8,18,32,10,2
(4)	104	Rf	Rutherfordium	2,8,18,32,32,10,2
(5)	23	V	Vanadium	2,8,11,2
(5)	41	Nb	Niobium	2,8,18,12,1
(5)	73	Ta	Tantalum	2,8,18,32,11,2
(5)	105	Db	Dubnium	2,8,18,32,32,11,2
(6)	24	Cr	Chromium	2,8,13,1
(6)	42	Mo	Molybdenum	2,8,18,13,1
(6)	74	W	Tungsten	2,8,18,32,12,2
(6)	106	Sg	Seaborgium	2,8,18,32,32,12,2
(7)	25	Mn	Manganese	2,8,13,2
(7)	43	Tc	Technetium	2,8,18,13,2
(7)	75	Re	Rhenium	2,8,18,32,13,2
(7)	107	Bh	Bohrium	2,8,18,32,32,13,2
(8)	26	Fe	Iron	2,8,14,2
(8)	44	Ru	Ruthenium	2,8,18,15,1
(8)	76	Os	Osmium	2,8,18,32,14,2
(8)	108	Hs	Hassium	2,8,18,32,32,14,2
(9)	27	Co	Cobalt	2,8,15,2
(9)	45	Rh	Rhodium	2,8,18,16,1
(9)	77	Ir	Iridium	2,8,18,32,15,2
(9)	109	Mt	Meitnerium	2,8,18,32,32,15,2
(10)	28	Ni	Nickel	2,8,16,2
(10)	46	Pd	Palladium	2,8,18,18,0
(10)	78	Pt	Platinum	2,8,18,32,17,1
(10)	110	Ds	Darmstadtium	2,8,18,32,32,17,1
(11)	29	Cu	Copper	2,8,18,1
(11)	47	Ag	Silver	2,8,18,18,1
(11)	79	Au	Gold	2,8,18,32,18,1
(11)	111	Rg	Roentgenium	2,8,18,32,32,18,1
(12)	30	Zn	Zinc	2,8,18,2
(12)	48	Cd	Cadmium	2,8,18,18,2
(12)	80	Hg	Mercury	2,8,18,32,18,2
(12)	112	Cn	Copernicium	2,8,18,32,32,18,2

Group 3 (13)

Atomic number	Symbol	Name	Electron arrangement
5	B	Boron	2,3
13	Al	Aluminium	2,8,3
31	Ga	Gallium	2,8,18,3
49	In	Indium	2,8,18,18,3
81	Tl	Thallium	2,8,18,32,18,3

Group 4 (14)

Atomic number	Symbol	Name	Electron arrangement
6	C	Carbon	2,4
14	Si	Silicon	2,8,4
32	Ge	Germanium	2,8,18,4
50	Sn	Tin	2,8,18,18,4
82	Pb	Lead	2,8,18,32,18,4

Group 5 (15)

Atomic number	Symbol	Name	Electron arrangement
7	N	Nitrogen	2,5
15	P	Phosphorus	2,8,5
33	As	Arsenic	2,8,18,5
51	Sb	Antimony	2,8,18,18,5
83	Bi	Bismuth	2,8,18,32,18,5

Group 6 (16)

Atomic number	Symbol	Name	Electron arrangement
8	O	Oxygen	2,6
16	S	Sulfur	2,8,6
34	Se	Selenium	2,8,18,6
52	Te	Tellurium	2,8,18,18,6
84	Po	Polonium	2,8,18,32,18,6

Group 7 (17)

Atomic number	Symbol	Name	Electron arrangement
9	F	Fluorine	2,7
17	Cl	Chlorine	2,8,7
35	Br	Bromine	2,8,18,7
53	I	Iodine	2,8,18,18,7
85	At	Astatine	2,8,18,32,18,7

Group 0 (18)

Atomic number	Symbol	Name	Electron arrangement
2	He	Helium	2
10	Ne	Neon	2,8
18	Ar	Argon	2,8,8
36	Kr	Krypton	2,8,18,8
54	Xe	Xenon	2,8,18,18,8
86	Rn	Radon	2,8,18,32,18,8

Lanthanides

Atomic number	Symbol	Name	Electron arrangement
57	La	Lanthanum	2,8,18,18,9,2
58	Ce	Cerium	2,8,18,20,8,2
59	Pr	Praseodymium	2,8,18,21,8,2
60	Nd	Neodymium	2,8,18,22,8,2
61	Pm	Promethium	2,8,18,23,8,2
62	Sm	Samarium	2,8,18,24,8,2
63	Eu	Europium	2,8,18,25,8,2
64	Gd	Gadolinium	2,8,18,25,9,2
65	Tb	Terbium	2,8,18,27,8,2
66	Dy	Dysprosium	2,8,18,28,8,2
67	Ho	Holmium	2,8,18,29,8,2
68	Er	Erbium	2,8,18,30,8,2
69	Tm	Thulium	2,8,18,31,8,2
70	Yb	Ytterbium	2,8,18,32,8,2
71	Lu	Lutetium	2,8,18,32,9,2

Actinides

Atomic number	Symbol	Name	Electron arrangement
89	Ac	Actinium	2,8,18,32,18,9,2
90	Th	Thorium	2,8,18,32,18,10,2
91	Pa	Protactinium	2,8,18,32,20,9,2
92	U	Uranium	2,8,18,32,21,9,2
93	Np	Neptunium	2,8,18,32,22,9,2
94	Pu	Plutonium	2,8,18,32,24,8,2
95	Am	Americium	2,8,18,32,25,8,2
96	Cm	Curium	2,8,18,32,25,9,2
97	Bk	Berkelium	2,8,18,32,27,8,2
98	Cf	Californium	2,8,18,32,28,8,2
99	Es	Einsteinium	2,8,18,32,29,8,2
100	Fm	Fermium	2,8,18,32,30,8,2
101	Md	Mendelevium	2,8,18,32,31,8,2
102	No	Nobelium	2,8,18,32,32,8,2
103	Lr	Lawrencium	2,8,18,32,32,9,2

FOR OFFICIAL USE

N5

National Qualifications 2014

Mark

X757/75/01

Physics
Section 1—Answer Grid and Section 2

THURSDAY, 22 MAY

9:00 AM – 11:00 AM

Fill in these boxes and read what is printed below.

Full name of centre

Town

Forename(s)

Surname

Number of seat

Date of birth

Day	Month	Year
D D	M M	Y Y

Scottish candidate number

Total marks — 110

SECTION 1 — 20 marks
Attempt ALL questions in this section.
Instructions for the completion of Section 1 are given on *Page two*.

SECTION 2 — 90 marks
Attempt ALL questions in this section.

Write your answers clearly in the spaces provided in this booklet. Additional space for answers and rough work is provided at the end of this booklet. If you use this space you must clearly identify the question number you are attempting. Any rough work must be written in this booklet. You should score through your rough work when you have written your final copy.

Use **blue** or **black** ink.

Reference may be made to the Data Sheet on *Page two* of the question paper X757/75/02 and to the Relationship Sheet X757/75/11.

Care should be taken to give an appropriate number of significant figures in the final answers to calculations.

Before leaving the examination room you must give this booklet to the Invigilator; if you do not, you may lose all the marks for this paper.

SECTION 1 — 20 marks

The questions for Section 1 are contained in the question paper X757/75/02.
Read these and record your answers on the answer grid on *Page three* opposite.
Do NOT use gel pens.

1. The answer to each question is **either** A, B, C, D or E. Decide what your answer is, then fill in the appropriate bubble (see sample question below).

2. There is **only one correct** answer to each question.

3. Any rough work must be written in the additional space for answers and rough work at the end of this booklet.

Sample Question

The energy unit measured by the electricity meter in your home is the:

    A   ampere

    B   kilowatt-hour

    C   watt

    D   coulomb

    E   volt.

The correct answer is **B**—kilowatt-hour.   The answer **B** bubble has been clearly filled in (see below).

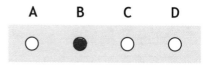

Changing an answer

If you decide to change your answer, cancel your first answer by putting a cross through it (see below) and fill in the answer you want. The answer below has been changed to **D**.

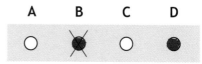

If you then decide to change back to an answer you have already scored out, put a tick (✓) to the **right** of the answer you want, as shown below:

                               or

SECTION 1 — Answer Grid

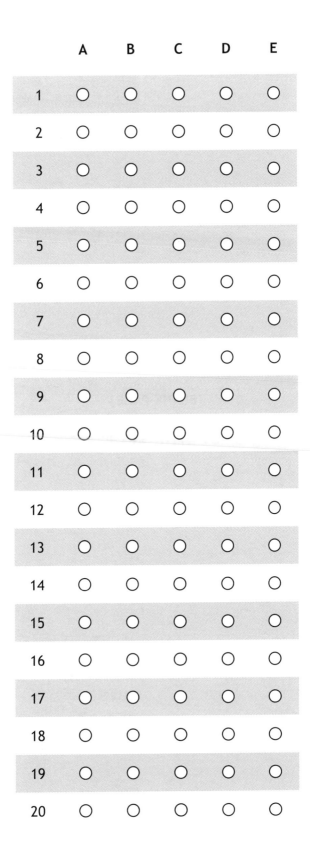

[Turn over

[BLANK PAGE]

DO NOT WRITE ON THIS PAGE

[Turn over for Question 1 on *Page six*

DO NOT WRITE ON THIS PAGE

SECTION 2 — 90 marks

Attempt ALL questions

1. A toy car contains an electric circuit which consists of a 12·0 V battery, an electric motor and two lamps.

The circuit diagram is shown.

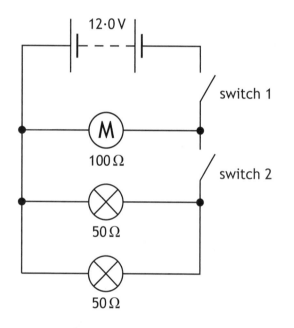

12·0 V

switch 1

M

100 Ω

switch 2

50 Ω

50 Ω

(a) Switch 1 is now closed.

Calculate the power dissipated in the motor when operating.

3

Space for working and answer

MARKS | DO NOT WRITE IN THIS MARGIN

1. (continued)

(b) Switch 2 is now also closed.

(i) Calculate the total resistance of the motor and the two lamps.   3

Space for working and answer

(ii) One of the lamps now develops a fault and stops working.

State the effect this has on the other lamp.

You **must** justify your answer.   2

Total marks   8

[Turn over

MARKS | DO NOT WRITE IN THIS MARGIN

2. A thermistor is used as a temperature sensor in a circuit to monitor and control the temperature of water in a tank. Part of the circuit is shown.

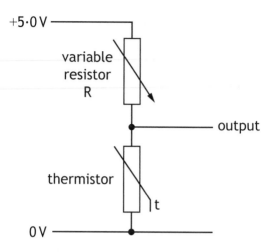

(a) (i) The variable resistor R is set at a resistance of 1050 Ω.

Calculate the resistance of the thermistor when the voltage across the thermistor is 2·0 V.

4

Space for working and answer

MARKS | DO NOT WRITE IN THIS MARGIN

2. (a) (continued)

(ii) The graph shows how the resistance of the thermistor varies with temperature.

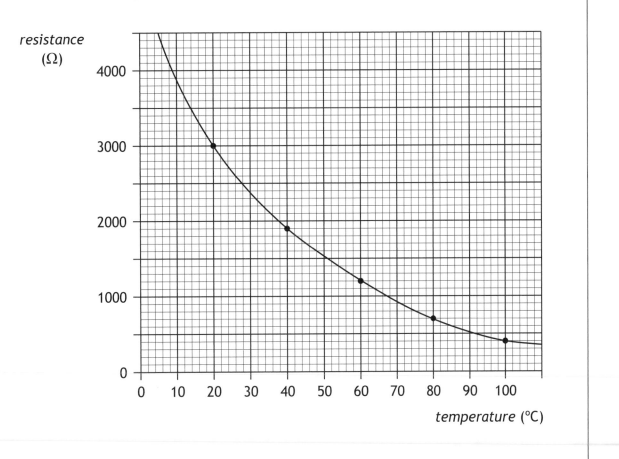

Use the graph to determine the temperature of the water when the voltage across the thermistor is 2·0 V.

1

2. **(continued)**

(b) The circuit is now connected to a switching circuit to operate a heater.

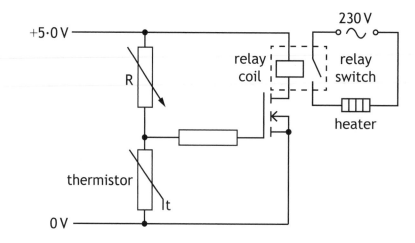

(i) Explain how the circuit operates to switch on the heater when the temperature falls below a certain value. **3**

(ii) The resistance of the variable resistor R is now increased.

What effect does this have on the temperature at which the heater is switched on?

You **must** justify your answer. **3**

Total marks 11

[Turn over for Question 3 on *Page twelve*

DO NOT WRITE ON THIS PAGE

MARKS | DO NOT WRITE IN THIS MARGIN

3. A student is investigating the specific heat capacity of three metal blocks X, Y and Z.

Each block has a mass of 1·0 kg.

A heater and thermometer are inserted into a block as shown.

The heater has a power rating of 15 W.

The initial temperature of the block is measured.

The heater is switched on for 10 minutes and the final temperature of the block is recorded.

This procedure is repeated for the other two blocks.

The student's results are shown in the table.

Block	Initial temperature (°C)	Final temperature (°C)
X	15	25
Y	15	85
Z	15	34

(a) Show that the energy provided by the heater to each block is 9000 J.   2

Space for working and answer

MARKS | DO NOT WRITE IN THIS MARGIN

3.   (continued)

(b) (i)   By referring to the results in the table, identify the block that has the greatest specific heat capacity.   1

(ii)   Calculate the specific heat capacity of the block identified in (b)(i).   3

Space for working and answer

(c) Due to energy losses, the specific heat capacities calculated in this investigation are different from the accepted values.

The student decides to improve the set up in order to obtain a value closer to the accepted value for each block.

(i)   Suggest a possible improvement that would reduce energy losses.   1

(ii)   State the effect that this improvement would have on the final temperature.   1

Total marks   8

MARKS

4. A student, fishing from a pier, counts four waves passing the end of the pier in 20 seconds. The student estimates that the wavelength of the waves is 12 m.

Not to scale

  (a) Calculate the speed of the water waves.   **4**

     Space for working and answer

MARKS | DO NOT WRITE IN THIS MARGIN

4. **(continued)**

(b) When looking down into the calm water behind the pier the student sees a fish.

Complete the diagram to show the path of a ray of light from the fish to the student.

You should include the normal in your diagram.　　3

(An additional diagram, if required, can be found on *Page thirty-one*.)

Total marks　7

[Turn over

5. The UV Index is an international standard measurement of the intensity of ultraviolet radiation from the Sun. Its purpose is to help people to effectively protect themselves from UV rays.

The UV index table is shown.

UV Index	Description
0 – 2	Low risk from the Sun's UV rays for the average person
3 – 5	Moderate risk of harm from unprotected Sun exposure
6 – 7	High risk of harm from unprotected Sun exposure
8 – 10	Very high risk of harm from unprotected Sun exposure
11+	Extreme risk of harm from unprotected Sun exposure

The UV index can be calculated using

$$UV\ index = \left[\begin{array}{c} total\ effect\ of \\ UV\ radiation \end{array} \times \begin{array}{c} elevation\ above \\ sea\ level\ adjustment \end{array} \times \begin{array}{c} cloud \\ adjustment \end{array} \right] \div 25$$

The UV index is then rounded to the nearest whole number.

The tables below give information for elevation above sea level and cloud cover.

Elevation above sea level (km)	Elevation above sea level adjustment
1	1·06
2	1·12
3	1·18

Cloud cover	Cloud adjustment
Clear skies	1·00
Scattered clouds	0·89
Broken clouds	0·73
Overcast skies	0·31

5. (continued)

MARKS

(a) At a particular location the total effect of UV radiation is 280.

The elevation is 2 km above sea level with overcast skies.

Calculate the UV index value for this location.

Space for working and answer

2

(b) Applying sunscreen to the skin is one method of protecting people from the Sun's harmful UV rays. UV radiation can be divided into three wavelength ranges, called UVA, UVB and UVC.

A manufacturer carries out some tests on experimental sunscreens P, Q and R to determine how effective they are at absorbing UV radiation. The test results are displayed in the graph.

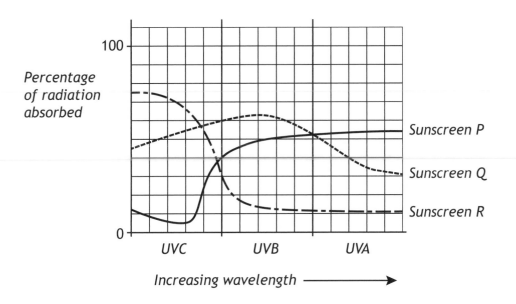

Using information from the graph, complete the following table.

2

	UVA	UVB	UVC
Type of sunscreen that absorbs most of this radiation		Sunscreen Q	
Type of sunscreen that absorbs least of this radiation	Sunscreen R		

(c) State one useful application of UV radiation.

1

Total marks 5

[Turn over

MARKS | DO NOT WRITE IN THIS MARGIN

6. A technician carries out an experiment, using the apparatus shown, to determine the half-life of a radioactive source.

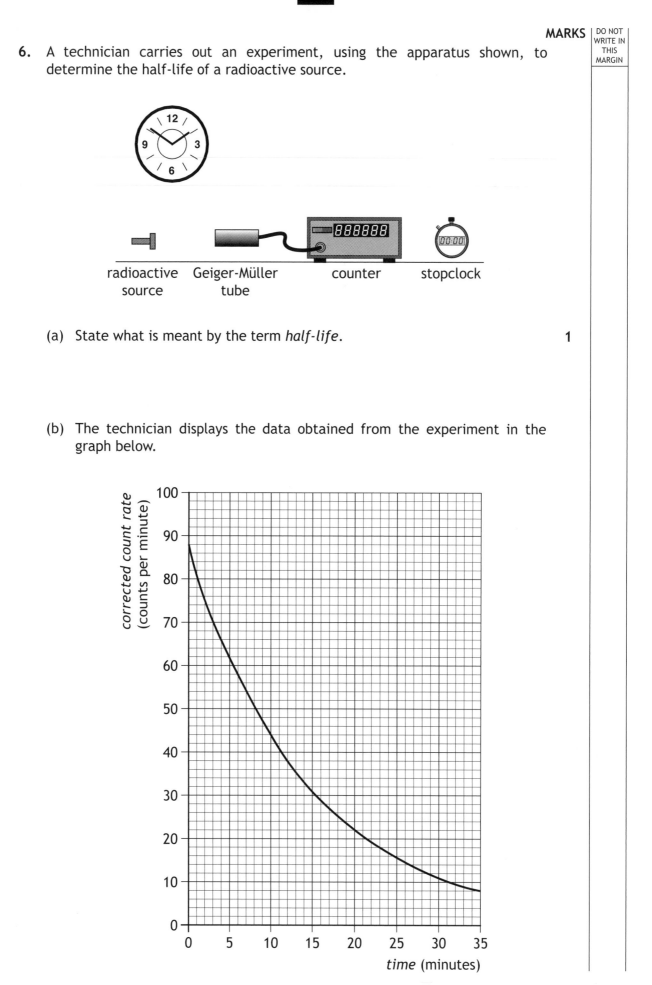

radioactive source   Geiger-Müller tube   counter   stopclock

(a) State what is meant by the term *half-life*.   1

(b) The technician displays the data obtained from the experiment in the graph below.

MARKS | DO NOT WRITE IN THIS MARGIN

6. (b) (continued)

(i) Describe how the apparatus could be used to obtain the experimental data required to produce this graph.

3

(ii) Use information from the graph to determine the half-life of the radioactive source.

1

(iii) Determine the corrected count rate after 40 minutes.

Space for working and answer

2

Total marks 7

[Turn over

7. A fire engine on its way to an emergency is travelling along a main street. The siren on the fire engine is sounding.

A student standing in a nearby street cannot see the fire engine but can hear the siren.

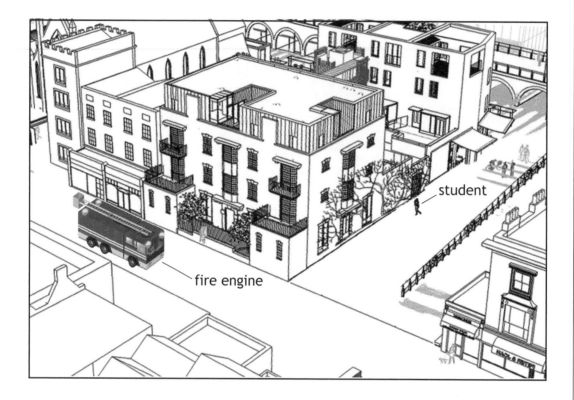

student

fire engine

Use your knowledge of physics to comment on why the student can hear the siren even though the fire engine is not in view.   3

8. An airport worker passes suitcases through an X-ray machine.

bibiphoto/shutterstock.com

(a) The worker has a mass of 80·0 kg and on a particular day absorbs 7·2 mJ of energy from the X-ray machine.

    (i) Calculate the absorbed dose received by the worker.   **3**

      Space for working and answer

    (ii) Calculate the equivalent dose received by the worker.   **3**

      Space for working and answer

MARKS | DO NOT WRITE IN THIS MARGIN

8. **(continued)**

(b) X-rays can cause ionisation.

Explain what is meant by *ionisation*.   1

Total marks   7

MARKS | DO NOT WRITE IN THIS MARGIN

9. A communications satellite is used to transmit live television broadcasts from the UK to Canada.

A student states that, to allow the live television broadcasts to be received in Canada, it is important that the satellite does not move.

Use your knowledge of physics to comment on this statement.   3

[Turn over

MARKS | DO NOT WRITE IN THIS MARGIN

10. In a rowing event a boat moves off in a straight line.

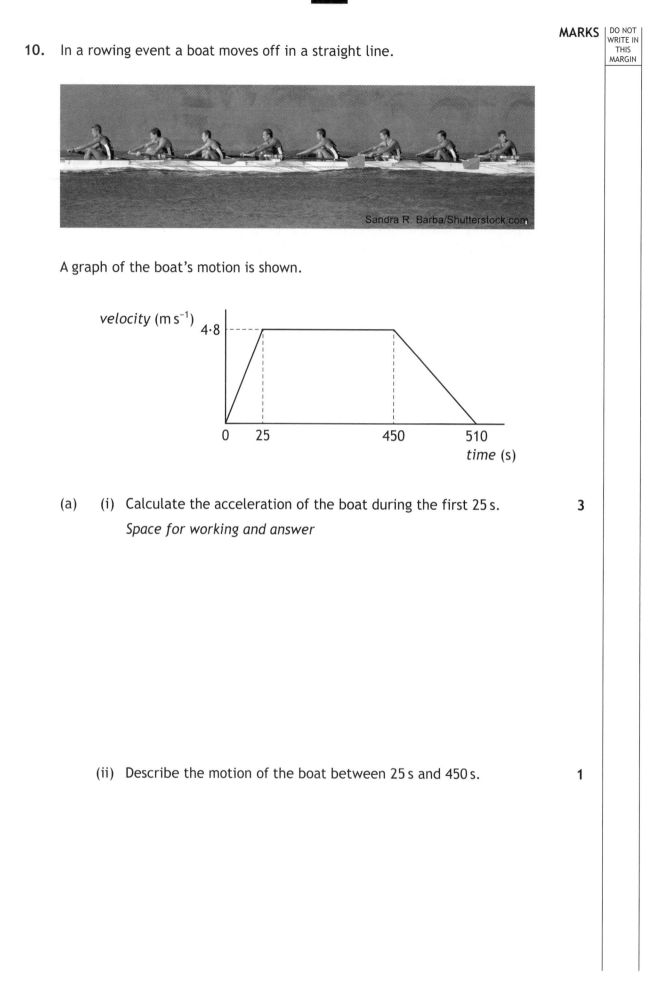

Sandra R. Barba/Shutterstock.com

A graph of the boat's motion is shown.

velocity (m s^{-1})

4·8

0   25   450   510

time (s)

(a) (i) Calculate the acceleration of the boat during the first 25 s.

Space for working and answer

3

(ii) Describe the motion of the boat between 25 s and 450 s.

1

MARKS | DO NOT WRITE IN THIS MARGIN

10. **(a)** **(continued)**

(iii) Draw a diagram showing the horizontal forces acting on the boat between 25 s and 450 s.

You **must** name these forces and show their directions.   **2**

(b) The boat comes to rest after 510 s.

(i) Calculate the total distance travelled by the boat.   **3**

Space for working and answer

(ii) Calculate the average velocity of the boat.

A direction is not required.   **3**

Space for working and answer

Total marks 12

MARKS | DO NOT WRITE IN THIS MARGIN

11. A helicopter is used to take tourists on sightseeing flights.

Information about the helicopter is shown in the table.

weight of empty helicopter	13 500 N
maximum take-off weight	24 000 N
cruising speed	$67\,\mathrm{m\,s^{-1}}$
maximum speed	$80\,\mathrm{m\,s^{-1}}$
maximum range	610 km

(a) The pilot and passengers are weighed before they board the helicopter.

Explain the reason for this.   **1**

(b) Six passengers and the pilot with a combined weight of 6125 N board the helicopter.

Determine the minimum upward force required by the helicopter at take-off.   **1**

Space for working and answer

MARKS | DO NOT WRITE IN THIS MARGIN

11. (continued)

(c) The helicopter travels 201 km at its cruising speed.

Calculate the time taken to travel this distance. **3**

Space for working and answer

Total marks 5

[Turn over

MARKS | DO NOT WRITE IN THIS MARGIN

12. A student is investigating the motion of water rockets. The water rocket is made from an upturned plastic bottle containing some water. Air is pumped into the bottle. When the pressure of the air is great enough the plastic bottle is launched upwards.

The mass of the rocket before launch is 0·94 kg.

(a) Calculate the weight of the water rocket. **3**

Space for working and answer

(b) Before launch, the water rocket rests on three fins on the ground.

The area of each fin in contact with the ground is $2·0 \times 10^{-4}\, m^2$.

Calculate the total pressure exerted on the ground by the fins. **4**

Space for working and answer

MARKS | DO NOT WRITE IN THIS MARGIN

12. **(continued)**

(c) Use Newton's Third Law to explain how the rocket launches.   **1**

(d) At launch, the initial upward thrust on the rocket is 370 N.

Calculate the initial acceleration of the rocket.   **4**

Space for working and answer

(e) The student launches the rocket a second time.

For this launch, the student adds a greater volume of water than before.

The same initial upward thrust acts on the rocket but it fails to reach the same height.

Explain why the rocket fails to reach the same height.   **2**

Total marks **14**

[END OF QUESTION PAPER]

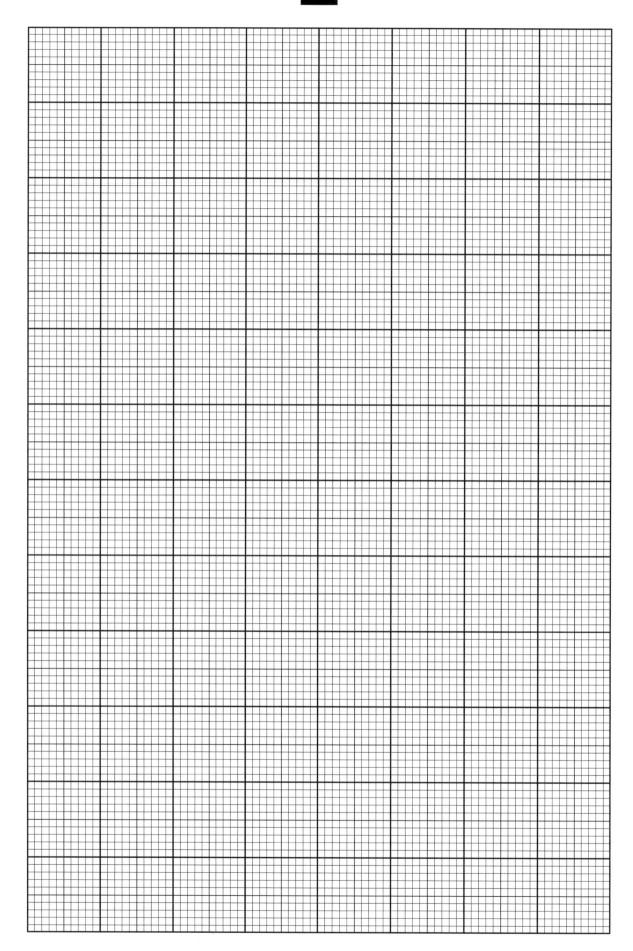

ADDITIONAL SPACE FOR ANSWERS AND ROUGH WORK

Additional diagram for Question 4 (b)

MARKS DO NOT WRITE IN THIS MARGIN

ADDITIONAL SPACE FOR ANSWERS AND ROUGH WORK

ADDITIONAL SPACE FOR ANSWERS AND ROUGH WORK

MARKS | DO NOT WRITE IN THIS MARGIN

[BLANK PAGE]

DO NOT WRITE ON THIS PAGE

NATIONAL 5

2015

National Qualifications 2015

X757/75/02

Physics
Section 1—Questions

TUESDAY, 5 MAY

9:00 AM – 11:00 AM

Instructions for the completion of Section 1 are given on *Page two* of your question and answer booklet X757/75/01.

Record your answers on the answer grid on *Page three* of your question and answer booklet.

Reference may be made to the Data Sheet on *Page two* of this booklet and to the Relationship Sheet X757/75/11.

Before leaving the examination room you must give your question and answer booklet to the Invigilator; if you do not, you may lose all the marks for this paper.

DATA SHEET

Speed of light in materials

Material	Speed in $m\,s^{-1}$
Air	3.0×10^8
Carbon dioxide	3.0×10^8
Diamond	1.2×10^8
Glass	2.0×10^8
Glycerol	2.1×10^8
Water	2.3×10^8

Gravitational field strengths

	Gravitational field strength on the surface in $N\,kg^{-1}$
Earth	9.8
Jupiter	23
Mars	3.7
Mercury	3.7
Moon	1.6
Neptune	11
Saturn	9.0
Sun	270
Uranus	8.7
Venus	8.9

Specific latent heat of fusion of materials

Material	Specific latent heat of fusion in $J\,kg^{-1}$
Alcohol	0.99×10^5
Aluminium	3.95×10^5
Carbon Dioxide	1.80×10^5
Copper	2.05×10^5
Iron	2.67×10^5
Lead	0.25×10^5
Water	3.34×10^5

Specific latent heat of vaporisation of materials

Material	Specific latent heat of vaporisation in $J\,kg^{-1}$
Alcohol	11.2×10^5
Carbon Dioxide	3.77×10^5
Glycerol	8.30×10^5
Turpentine	2.90×10^5
Water	22.6×10^5

Speed of sound in materials

Material	Speed in $m\,s^{-1}$
Aluminium	5200
Air	340
Bone	4100
Carbon dioxide	270
Glycerol	1900
Muscle	1600
Steel	5200
Tissue	1500
Water	1500

Specific heat capacity of materials

Material	Specific heat capacity in $J\,kg^{-1}\,°C^{-1}$
Alcohol	2350
Aluminium	902
Copper	386
Glass	500
Ice	2100
Iron	480
Lead	128
Oil	2130
Water	4180

Melting and boiling points of materials

Material	Melting point in °C	Boiling point in °C
Alcohol	−98	65
Aluminium	660	2470
Copper	1077	2567
Glycerol	18	290
Lead	328	1737
Iron	1537	2737

Radiation weighting factors

Type of radiation	Radiation weighting factor
alpha	20
beta	1
fast neutrons	10
gamma	1
slow neutrons	3
X-rays	1

SECTION 1

Attempt ALL questions

1. Two circuits are set up as shown.

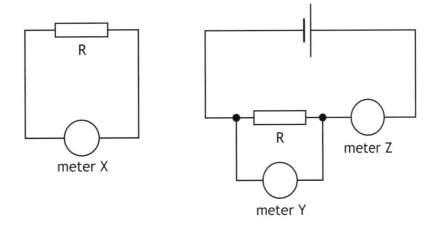

Both circuits are used to determine the resistance of resistor R.

Which row in the table identifies meter X, meter Y and meter Z?

	meter X	meter Y	meter Z
A	ohmmeter	voltmeter	ammeter
B	ohmmeter	ammeter	voltmeter
C	voltmeter	ammeter	ohmmeter
D	ammeter	voltmeter	ohmmeter
E	voltmeter	ohmmeter	ammeter

2. Which of the following statements is/are correct?

    I   The voltage of a battery is the number of joules of energy it gives to each coulomb of charge.

    II   A battery only has a voltage when it is connected in a complete circuit.

    III Electrons are free to move within an insulator.

    A   I only

    B   II only

    C   III only

    D   II and III only

    E   I, II and III

[Turn over

3.   A circuit is set up as shown.

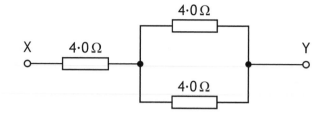

The resistance between X and Y is

A $1 \cdot 3\,\Omega$

B $4 \cdot 5\,\Omega$

C $6 \cdot 0\,\Omega$

D $8 \cdot 0\,\Omega$

E $12\,\Omega$.

4.   The rating plate on an electrical appliance is shown.

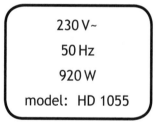

The resistance of this appliance is

A $0 \cdot 017\,\Omega$

B $0 \cdot 25\,\Omega$

C $4 \cdot 0\,\Omega$

D $18 \cdot 4\,\Omega$

E $57 \cdot 5\,\Omega$.

5. A syringe containing air is sealed at one end as shown.

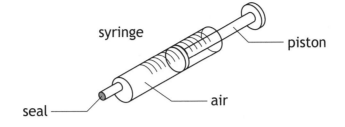

The piston is pushed in slowly.

There is no change in temperature of the air inside the syringe.

Which of the following statements describes and explains the change in pressure of the air in the syringe?

A   The pressure increases because the air particles have more kinetic energy.

B   The pressure increases because the air particles hit the sides of the syringe more frequently.

C   The pressure increases because the air particles hit the sides of the syringe less frequently.

D   The pressure decreases because the air particles hit the sides of the syringe with less force.

E   The pressure decreases because the air particles have less kinetic energy.

6. The pressure of a fixed mass of gas is 150 kPa at a temperature of 27 °C.

The temperature of the gas is now increased to 47 °C.

The volume of the gas remains constant.

The pressure of the gas is now

A   86 kPa

B   141 kPa

C   150 kPa

D   160 kPa

E   261 kPa.

[Turn over

7. The diagram represents a water wave.

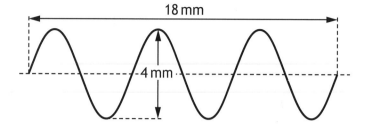

The wavelength of the water wave is

A   2 mm

B   3 mm

C   4 mm

D   6 mm

E   18 mm.

8. A student makes the following statements about different types of electromagnetic waves.

    I   Light waves are transverse waves.

    II   Radio waves travel at 340 m s^{-1} through air.

    III Ultraviolet waves have a longer wavelength than infrared waves.

    Which of these statements is/are correct?

    A   I only

    B   I and II only

    C   I and III only

    D   II and III only

    E   I, II and III

9. Alpha radiation ionises an atom.

    Which statement describes what happens to the atom?

    A   The atom splits in half.

    B   The atom releases a neutron.

    C   The atom becomes positively charged.

    D   The atom gives out gamma radiation.

    E   The atom releases heat.

10. A sample of tissue is irradiated using a radioactive source.

   A student makes the following statements.

   The equivalent dose received by the tissue is

   I   reduced by shielding the tissue with a lead screen

   II   increased as the distance from the source to the tissue is increased

   III increased by increasing the time of exposure of the tissue to the radiation.

   Which of the statements is/are correct?

   A   I only

   B   II only

   C   I and II only

   D   II and III only

   E   I and III only

11. A sample of tissue receives an absorbed dose of 16 µGy from alpha particles.

   The radiation weighting factor for alpha particles is 20.

   The equivalent dose received by the sample is

   A   0·80 µSv

   B   1·25 µSv

   C   4 µSv

   D   36 µSv

   E   320 µSv.

12. For a particular radioactive source, 240 atoms decay in 1 minute.

   The activity of this source is

   A   4 Bq

   B   180 Bq

   C   240 Bq

   D   300 Bq

   E   14 400 Bq.

[Turn over

13. The letters **X, Y** and **Z** represent missing words from the following passage.

During a nuclear ......X...... reaction two nuclei of smaller mass number combine to produce a nucleus of larger mass number. During a nuclear ....Y....... reaction a nucleus of larger mass number splits into two nuclei of smaller mass number. Both of these reactions are important because these processes can release .....Z...... .

Which row in the table shows the missing words?

	X	Y	Z
A	fusion	fission	electrons
B	fission	fusion	energy
C	fusion	fission	protons
D	fission	fusion	protons
E	fusion	fission	energy

14. Which of the following quantities is fully described by its magnitude?

 A Force

 B Displacement

 C Energy

 D Velocity

 E Acceleration

15. The table shows the velocities of three objects X, Y and Z over a period of 3 seconds. Each object is moving in a straight line.

Time (s)	0	1	2	3
Velocity of X (m s^{-1})	2	4	6	8
Velocity of Y (m s^{-1})	0	1	2	3
Velocity of Z (m s^{-1})	0	2	5	9

Which of the following statements is/are correct?

I   X moves with constant velocity.

II   Y moves with constant acceleration.

III   Z moves with constant acceleration.

A   I only

B   II only

C   I and II only

D   I and III only

E   II and III only

16. A car of mass 1200 kg is travelling along a straight level road at a constant speed of 20 m s^{-1}.

The driving force on the car is 2500 N. The frictional force on the car is 2500 N.

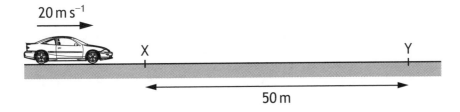

The work done moving the car between point X and point Y is

A   0 J

B   11 800 J

C   125 000 J

D   240 000 J

E   250 000 J.

[Turn over

17. A person sits on a chair which rests on the Earth. The person exerts a downward force on the chair.

Which of the following is the reaction to this force?

A   The force of the chair on the person

B   The force of the person on the chair

C   The force of the Earth on the person

D   The force of the chair on the Earth

E   The force of the person on the Earth

18. A package falls vertically from a helicopter. After some time the package reaches its terminal velocity.

A group of students make the following statements about the package when it reaches its terminal velocity.

I   The weight of the package is less than the air resistance acting on the package.

II   The forces acting on the package are balanced.

III The package is accelerating towards the ground at $9 \cdot 8 \, \text{m s}^{-2}$.

Which of these statements is/are correct?

A   I only

B   II only

C   III only

D   I and III only

E   II and III only

19. The distance from the Sun to Proxima Centauri is 4·3 light years.

    This distance is equivalent to

    A $1·4 \times 10^8$ m

    B $1·6 \times 10^{14}$ m

    C $6·8 \times 10^{14}$ m

    D $9·5 \times 10^{15}$ m

    E $4·1 \times 10^{16}$ m.

20. Light from a star is split into a line spectrum of different colours. The line spectrum from the star is shown, along with the line spectra of the elements calcium, helium, hydrogen and sodium.

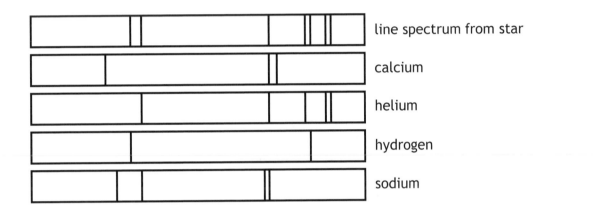

The elements present in this star are

    A sodium and calcium

    B calcium and helium

    C hydrogen and sodium

    D helium and hydrogen

    E calcium, sodium and hydrogen.

**[END OF SECTION 1. NOW ATTEMPT THE QUESTIONS IN SECTION 2
OF YOUR QUESTION AND ANSWER BOOKLET]**

[BLANK PAGE]

DO NOT WRITE ON THIS PAGE

National Qualifications 2015

X757/75/11

**Physics
Relationships Sheet**

TUESDAY, 5 MAY

9:00 AM – 11:00 AM

$E_p = mgh$

$E_k = \frac{1}{2}mv^2$

$Q = It$

$V = IR$

$R_T = R_1 + R_2 + \ldots$

$\frac{1}{R_T} = \frac{1}{R_1} + \frac{1}{R_2} + \ldots$

$V_2 = \left(\frac{R_2}{R_1 + R_2} \right) V_s$

$\frac{V_1}{V_2} = \frac{R_1}{R_2}$

$P = \frac{E}{t}$

$P = IV$

$P = I^2 R$

$P = \frac{V^2}{R}$

$E_h = cm\Delta T$

$p = \frac{F}{A}$

$\frac{pV}{T} = \text{constant}$

$p_1 V_1 = p_2 V_2$

$\frac{p_1}{T_1} = \frac{p_2}{T_2}$

$\frac{V_1}{T_1} = \frac{V_2}{T_2}$

$d = vt$

$v = f\lambda$

$T = \frac{1}{f}$

$A = \frac{N}{t}$

$D = \frac{E}{m}$

$H = Dw_R$

$\dot{H} = \frac{H}{t}$

$s = vt$

$d = \overline{v}t$

$s = \overline{v}t$

$a = \frac{v-u}{t}$

$W = mg$

$F = ma$

$E_w = Fd$

$E_h = ml$

Additional Relationships

Circle

circumference $= 2\pi r$

area $= \pi r^2$

Sphere

area $= 4\pi r^2$

volume $= \frac{4}{3}\pi r^3$

Trigonometry

$\sin\theta = \dfrac{\text{opposite}}{\text{hypotenuse}}$

$\cos\theta = \dfrac{\text{adjacent}}{\text{hypotenuse}}$

$\tan\theta = \dfrac{\text{opposite}}{\text{adjacent}}$

$\sin^2\theta + \cos^2\theta = 1$

Electron Arrangements of Elements

Key

| Atomic number |
| Symbol |
| Electron arrangement |
| Name |

Group 1 (1)

Atomic number	Symbol	Electron arrangement	Name
1	H	(1)	Hydrogen
3	Li	2,1	Lithium
11	Na	2,8,1	Sodium
19	K	2,8,8,1	Potassium
37	Rb	2,8,18,8,1	Rubidium
55	Cs	2,8,18,18,8,1	Caesium
87	Fr	2,8,18,32,18,8,1	Francium

Group 2 (2)

Atomic number	Symbol	Electron arrangement	Name
4	Be	2,2	Beryllium
12	Mg	2,8,2	Magnesium
20	Ca	2,8,8,2	Calcium
38	Sr	2,8,18,8,2	Strontium
56	Ba	2,8,18,18,8,2	Barium
88	Ra	2,8,18,32,18,8,2	Radium

Transition Elements

(3)
21	Sc	2,8,9,2	Scandium
39	Y	2,8,18,9,2	Yttrium
57	La	2,8,18,18,9,2	Lanthanum
89	Ac	2,8,18,32,18,9,2	Actinium

(4)
22	Ti	2,8,10,2	Titanium
40	Zr	2,8,18,10,2	Zirconium
72	Hf	2,8,18,32,10,2	Hafnium
104	Rf	2,8,18,32,32,10,2	Rutherfordium

(5)
23	V	2,8,11,2	Vanadium
41	Nb	2,8,18,12,1	Niobium
73	Ta	2,8,18,32,11,2	Tantalum
105	Db	2,8,18,32,32,11,2	Dubnium

(6)
24	Cr	2,8,13,1	Chromium
42	Mo	2,8,18,13,1	Molybdenum
74	W	2,8,18,32,12,2	Tungsten
106	Sg	2,8,18,32,32,12,2	Seaborgium

(7)
25	Mn	2,8,13,2	Manganese
43	Tc	2,8,18,13,2	Technetium
75	Re	2,8,18,32,13,2	Rhenium
107	Bh	2,8,18,32,32,13,2	Bohrium

(8)
26	Fe	2,8,14,2	Iron
44	Ru	2,8,18,15,1	Ruthenium
76	Os	2,8,18,32,14,2	Osmium
108	Hs	2,8,18,32,32,14,2	Hassium

(9)
27	Co	2,8,15,2	Cobalt
45	Rh	2,8,18,16,1	Rhodium
77	Ir	2,8,18,32,15,2	Iridium
109	Mt	2,8,18,32,32,15,2	Meitnerium

(10)
28	Ni	2,8,16,2	Nickel
46	Pd	2,8,18,18,0	Palladium
78	Pt	2,8,18,32,17,1	Platinum
110	Ds	2,8,18,32,32,17,1	Darmstadtium

(11)
29	Cu	2,8,18,1	Copper
47	Ag	2,8,18,18,1	Silver
79	Au	2,8,18,32,18,1	Gold
111	Rg	2,8,18,32,32,18,1	Roentgenium

(12)
30	Zn	2,8,18,2	Zinc
48	Cd	2,8,18,18,2	Cadmium
80	Hg	2,8,18,32,18,2	Mercury
112	Cn	2,8,18,32,32,18,2	Copernicium

Group 3 (13)

5	B	2,3	Boron
13	Al	2,8,3	Aluminium
31	Ga	2,8,18,3	Gallium
49	In	2,8,18,18,3	Indium
81	Tl	2,8,18,32,18,3	Thallium

Group 4 (14)

6	C	2,4	Carbon
14	Si	2,8,4	Silicon
32	Ge	2,8,18,4	Germanium
50	Sn	2,8,18,18,4	Tin
82	Pb	2,8,18,32,18,4	Lead

Group 5 (15)

7	N	2,5	Nitrogen
15	P	2,8,5	Phosphorus
33	As	2,8,18,5	Arsenic
51	Sb	2,8,18,18,5	Antimony
83	Bi	2,8,18,32,18,5	Bismuth

Group 6 (16)

8	O	2,6	Oxygen
16	S	2,8,6	Sulfur
34	Se	2,8,18,6	Selenium
52	Te	2,8,18,18,6	Tellurium
84	Po	2,8,18,32,18,6	Polonium

Group 7 (17)

9	F	2,7	Fluorine
17	Cl	2,8,7	Chlorine
35	Br	2,8,18,7	Bromine
53	I	2,8,18,18,7	Iodine
85	At	2,8,18,32,18,7	Astatine

Group 0 (18)

2	He	2	Helium
10	Ne	2,8	Neon
18	Ar	2,8,8	Argon
36	Kr	2,8,18,8	Krypton
54	Xe	2,8,18,18,8	Xenon
86	Rn	2,8,18,32,18,8	Radon

Lanthanides

Atomic number	Symbol	Electron arrangement	Name
57	La	2,8,18,18,9,2	Lanthanum
58	Ce	2,8,18,20,8,2	Cerium
59	Pr	2,8,18,21,8,2	Praseodymium
60	Nd	2,8,18,22,8,2	Neodymium
61	Pm	2,8,18,23,8,2	Promethium
62	Sm	2,8,18,24,8,2	Samarium
63	Eu	2,8,18,25,8,2	Europium
64	Gd	2,8,18,25,9,2	Gadolinium
65	Tb	2,8,18,27,8,2	Terbium
66	Dy	2,8,18,28,8,2	Dysprosium
67	Ho	2,8,18,29,8,2	Holmium
68	Er	2,8,18,30,8,2	Erbium
69	Tm	2,8,18,31,8,2	Thulium
70	Yb	2,8,18,32,8,2	Ytterbium
71	Lu	2,8,18,32,9,2	Lutetium

Actinides

Atomic number	Symbol	Electron arrangement	Name
89	Ac	2,8,18,32,18,9,2	Actinium
90	Th	2,8,18,32,18,10,2	Thorium
91	Pa	2,8,18,32,20,9,2	Protactinium
92	U	2,8,18,32,21,9,2	Uranium
93	Np	2,8,18,32,22,9,2	Neptunium
94	Pu	2,8,18,32,24,8,2	Plutonium
95	Am	2,8,18,32,25,8,2	Americium
96	Cm	2,8,18,32,25,9,2	Curium
97	Bk	2,8,18,32,27,8,2	Berkelium
98	Cf	2,8,18,32,28,8,2	Californium
99	Es	2,8,18,32,29,8,2	Einsteinium
100	Fm	2,8,18,32,30,8,2	Fermium
101	Md	2,8,18,32,31,8,2	Mendelevium
102	No	2,8,18,32,32,8,2	Nobelium
103	Lr	2,8,18,32,32,9,2	Lawrencium

N5

National Qualifications 2015

Mark

X757/75/01

**Physics
Section 1—Answer Grid
and Section 2**

TUESDAY, 5 MAY

9:00 AM – 11:00 AM

Fill in these boxes and read what is printed below.

Full name of centre

Town

Forename(s)

Surname

Number of seat

Date of birth

Day   Month   Year   Scottish candidate number

Total marks — 110

SECTION 1 — 20 marks
Attempt ALL questions.
Instructions for the completion of Section 1 are given on *Page two*.

SECTION 2 — 90 marks
Attempt ALL questions.

Reference may be made to the Data Sheet on *Page two* of the question paper X757/75/02 and to the Relationship Sheet X757/75/11.

Care should be taken to give an appropriate number of significant figures in the final answers to calculations.

Write your answers clearly in the spaces provided in this booklet. Additional space for answers and rough work is provided at the end of this booklet. If you use this space you must clearly identify the question number you are attempting. Any rough work must be written in this booklet. You should score through your rough work when you have written your final copy.

Use **blue** or **black** ink.

Before leaving the examination room you must give this booklet to the Invigilator; if you do not, you may lose all the marks for this paper.

SECTION 1 — 20 marks

The questions for Section 1 are contained in the question paper X757/75/02.
Read these and record your answers on the answer grid on *Page three* opposite.
Use **blue** or **black** ink. Do NOT use gel pens or pencil.

1. The answer to each question is **either** A, B, C, D or E. Decide what your answer is, then fill in the appropriate bubble (see sample question below).

2. There is **only one correct** answer to each question.

3. Any rough work must be written in the additional space for answers and rough work at the end of this booklet.

Sample Question

The energy unit measured by the electricity meter in your home is the:

A   ampere

B   kilowatt-hour

C   watt

D   coulomb

E   volt.

The correct answer is **B**—kilowatt-hour. The answer **B** bubble has been clearly filled in (see below).

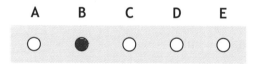

Changing an answer

If you decide to change your answer, cancel your first answer by putting a cross through it (see below) and fill in the answer you want. The answer below has been changed to **D**.

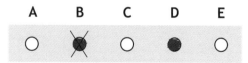

If you then decide to change back to an answer you have already scored out, put a tick (✓) to the **right** of the answer you want, as shown below:

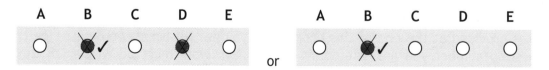

SECTION 1 — Answer Grid

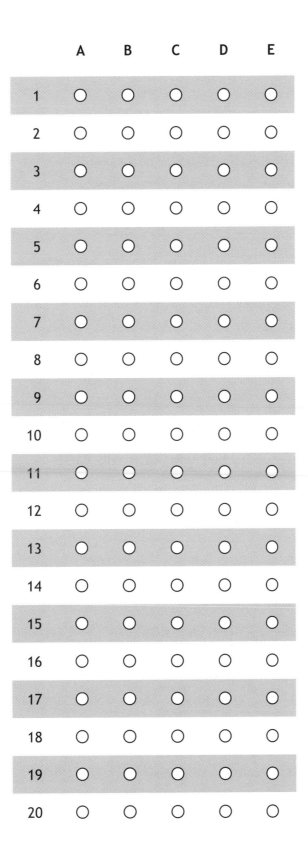

[Turn over

[BLANK PAGE]

DO NOT WRITE ON THIS PAGE

Page five

[Turn over for Question 1 on *Page six*

DO NOT WRITE ON THIS PAGE

SECTION 2 — 90 marks

Attempt ALL questions

1. A student sets up the following circuit using a battery, two lamps, a switch and a resistor.

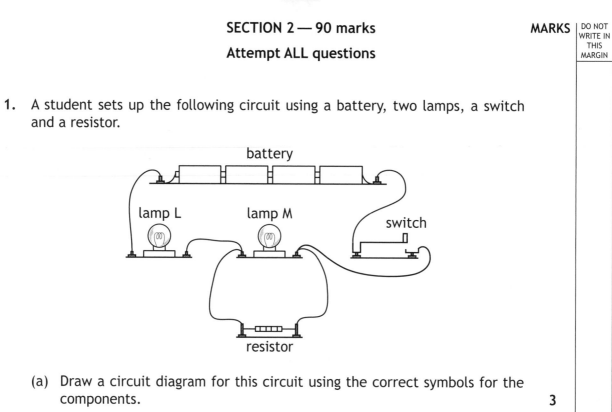

(a) Draw a circuit diagram for this circuit using the correct symbols for the components.

    3

(b) Each lamp is rated 2·5 V, 0·50 A.

    Calculate the resistance of one of the lamps when it is operating at the correct voltage.

    3

    Space for working and answer

MARKS DO NOT WRITE IN THIS MARGIN

1. **(continued)**

(c) When the switch is closed, will lamp L be brighter, dimmer or the same brightness as lamp M?

You **must** justify your answer. 3

[Turn over

MARKS | DO NOT WRITE IN THIS MARGIN

2. (a) A student investigates the electrical properties of three different components; a lamp, an LED and a fixed resistor.

Current-voltage graphs produced from the student's results are shown.

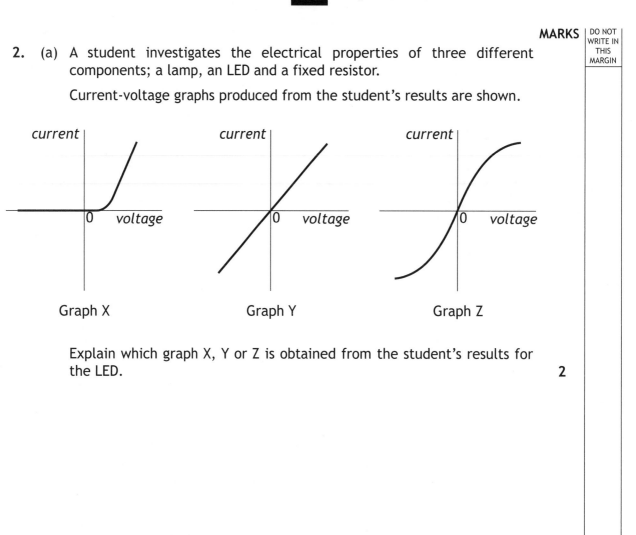

Graph X　　　　　Graph Y　　　　　Graph Z

Explain which graph X, Y or Z is obtained from the student's results for the LED.

2

(b) One of the components is operated at 4·0 V with a current of 0·50 A for 60 seconds.

(i) Calculate the energy transferred to the component during this time.

4

Space for working and answer

MARKS | DO NOT WRITE IN THIS MARGIN

2. (b) (continued)

(ii) Calculate the charge which passes through this component during this time.

3

Space for working and answer

[Turn over

3. A technician uses pulses of ultrasound (high frequency sound) to detect imperfections in a sample of steel.

The pulses of ultrasound are transmitted into the steel.

The speed of ultrasound in steel is $5200\,\text{m s}^{-1}$.

Where there are no imperfections, the pulses of ultrasound travel through the steel and are reflected by the back wall of the steel.

Where there are imperfections in the steel, the pulses of ultrasound are reflected by these imperfections.

The reflected pulses return through the sample and are detected by the ultrasound receiver.

The technician transmits pulses of ultrasound into the steel at positions X, Y and Z as shown.

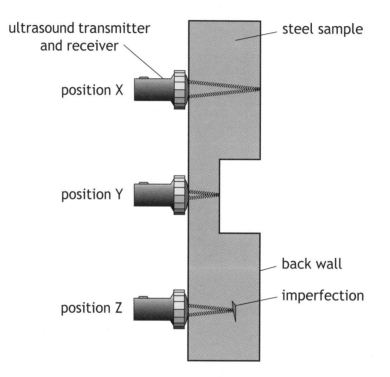

The times between the pulses being transmitted and received for positions X and Y are shown in the graph.

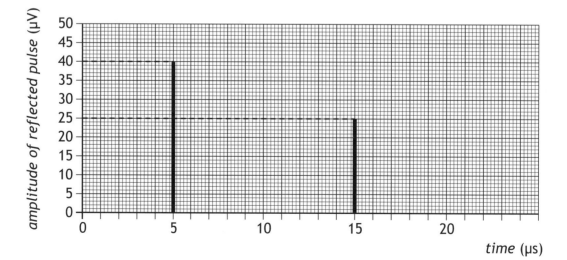

3. (continued)

(a) (i) State the time taken between the pulse being transmitted and received at position X.

1

(ii) Calculate the thickness of the steel sample at position X.

4

Space for working and answer

(b) On the graph on the previous page, draw a line to show the reflected pulse from position Z.

2

(c) The ultrasound pulses used have a period of 4·0 µs.

(i) Show that the frequency of the ultrasound pulses is $2·5 \times 10^5$ Hz.

2

Space for working and answer

(ii) Calculate the wavelength of the ultrasound pulses in the steel sample.

3

Space for working and answer

MARKS | DO NOT WRITE IN THIS MARGIN

3. (continued)

(d) The technician replaces the steel sample with a brass sample.

The brass sample has the same thickness as the steel sample at position X.

The technician transmits pulses of ultrasound into the brass at position P as shown.

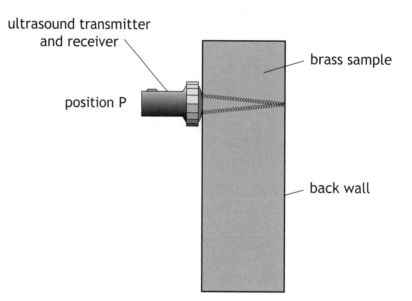

The time between the ultrasound pulse being transmitted and received at position P is greater than the time recorded at position X in the steel sample.

State whether the speed of ultrasound in brass is less than, equal to or greater than the speed of ultrasound in steel.

You **must** justify your answer.   2

MARKS | DO NOT WRITE IN THIS MARGIN

4. A science technician removes two metal blocks from an oven. Immediately after the blocks are removed from the oven the technician measures the temperature of each block, using an infrared thermometer. The temperature of each block is 230 °C.

After several minutes the temperature of each block is measured again. One block is now at a temperature of 123 °C and the other block is at a temperature of 187 °C.

Using your knowledge of physics, comment on possible explanations for this difference in temperature.

3

[Turn over

MARKS | DO NOT WRITE IN THIS MARGIN

5. Diamonds are popular and sought after gemstones.

Light is refracted as it enters and leaves a diamond.

The diagram shows a ray of light entering a diamond.

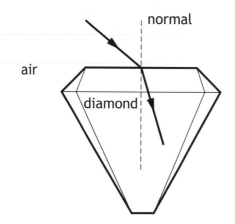

(a) On the diagram, label the angle of incidence i and the angle of refraction r.

1

(b) State what happens to the speed of the light as it enters the diamond.

1

(c) The optical density of a gemstone is a measure of its ability to refract light.

Gemstones of higher optical density cause more refraction.

A ray of light is directed into a gemstone at an angle of incidence of 45°.

The angle of refraction is then measured.

This is repeated for different gemstones.

Gemstone	Angle of refraction
A	24·3°
B	17·0°
C	27·3°
D	19·0°
E	25·5°

Diamond is known to have the highest optical density.

Identify which gemstone is most likely to be diamond.

1

MARKS

5. **(continued)**

(d) Diamond is one of the hardest known substances.

Synthetic diamonds are attached to the cutting edges of drill bits for use in the oil industry.

These drill bits are able to cut into rock.

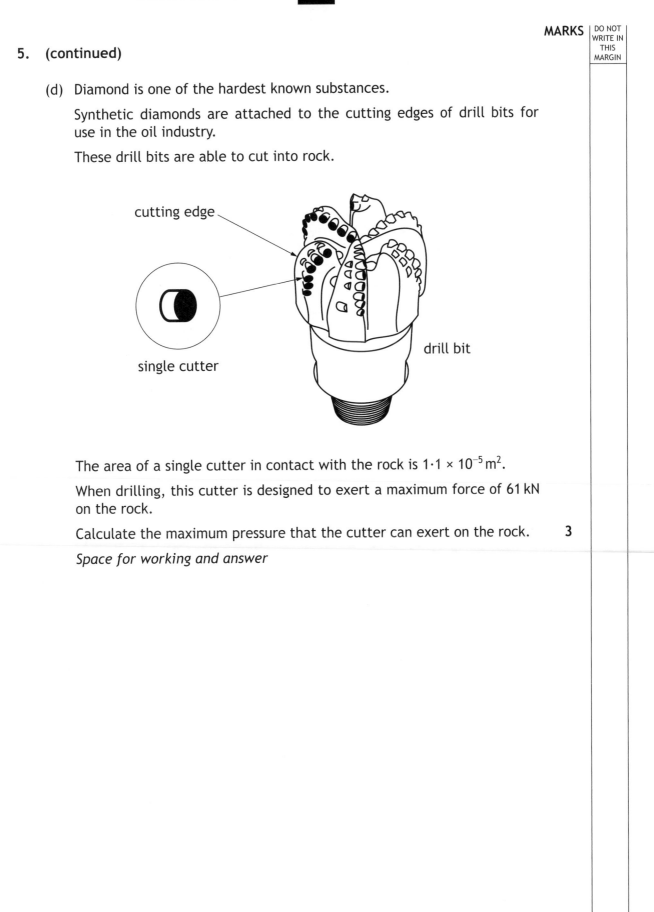

The area of a single cutter in contact with the rock is $1 \cdot 1 \times 10^{-5} \, \text{m}^2$.

When drilling, this cutter is designed to exert a maximum force of 61 kN on the rock.

Calculate the maximum pressure that the cutter can exert on the rock.   3

Space for working and answer

[Turn over

MARKS | DO NOT WRITE IN THIS MARGIN

6. A paper mill uses a radioactive source in a system to monitor the thickness of paper.

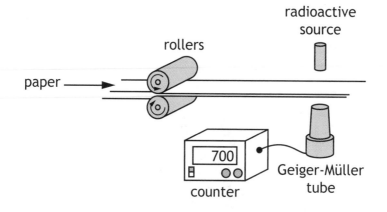

Radiation passing through the paper is detected by the Geiger-Müller tube. The count rate is displayed on the counter as shown. The radioactive source has a half-life that allows the system to run continuously.

(a) State what happens to the count rate if the thickness of the paper decreases.

1

(b) The following radioactive sources are available.

Radioactive Source	Half-life	Radiation emitted
W	600 years	alpha
X	50 years	beta
Y	4 hours	beta
Z	350 years	gamma

(i) State which radioactive source should be used.

You **must** explain your answer.

3

MARKS | DO NOT WRITE IN THIS MARGIN

6. **(b)** **(continued)**

(ii) State what is meant by the term *half-life*. 1

(iii) State what is meant by a gamma ray. 1

(c) The graph below shows how the activity of another radioactive source varies with time.

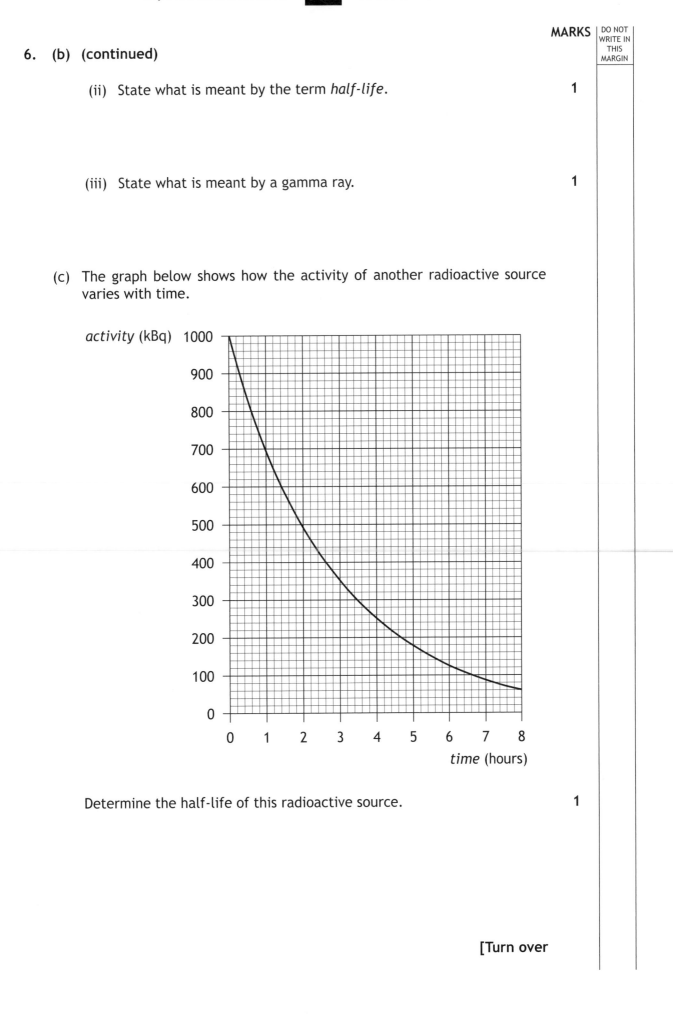

Determine the half-life of this radioactive source. 1

[Turn over

MARKS | DO NOT WRITE IN THIS MARGIN

7. A ship of mass $5 \cdot 0 \times 10^{6}$ kg leaves a port. Its engine produces a forward force of $8 \cdot 0 \times 10^{3}$ N. A tugboat pushes against one side of the ship as shown. The tugboat applies a pushing force of $6 \cdot 0 \times 10^{3}$ N.

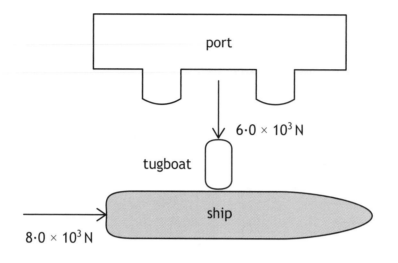

(a) (i) By scale drawing, or otherwise, determine the size of the resultant force acting on the ship.

Space for working and answer

2

(ii) Determine the direction of the resultant force relative to the $8 \cdot 0 \times 10^{3}$ N force.

Space for working and answer

2

MARKS

7. (a) (continued)

(iii) Calculate the size of the acceleration of the ship. **3**

Space for working and answer

(b) Out in the open sea the ship comes to rest.

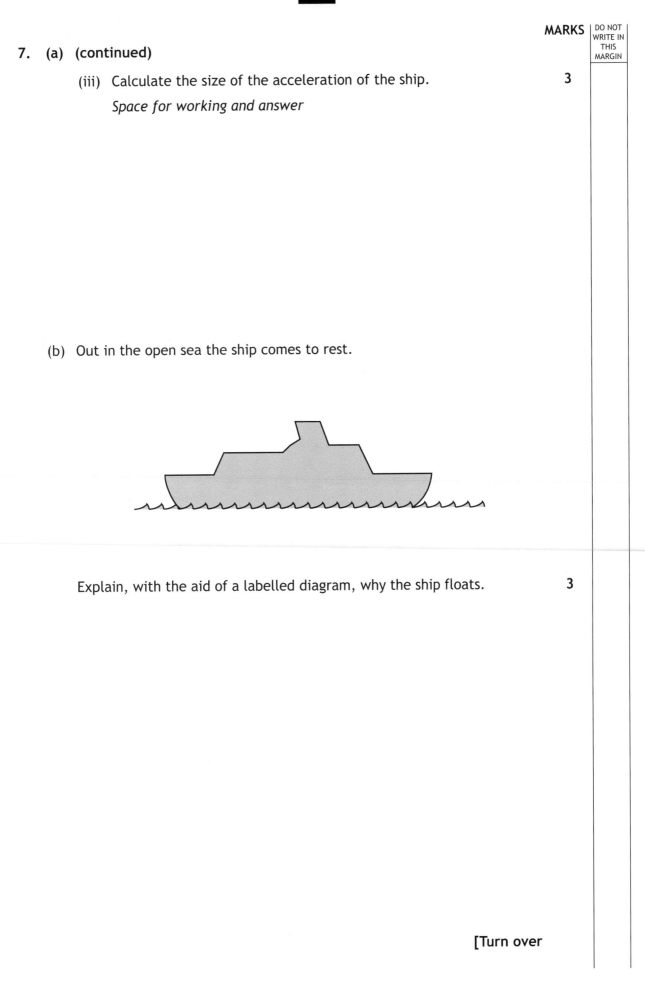

Explain, with the aid of a labelled diagram, why the ship floats. **3**

[Turn over

MARKS | DO NOT WRITE IN THIS MARGIN

8. A student is investigating the motion of a trolley down a ramp.

(a) The student uses the apparatus shown to carry out an experiment to determine the acceleration of a trolley as it rolls down a ramp.

The trolley is released from rest at the top of the ramp.

(i) State the measurements the student must make to calculate the acceleration of the trolley.

3

(ii) Suggest one reason why the acceleration calculated from these measurements might not be accurate.

1

MARKS | DO NOT WRITE IN THIS MARGIN

8. **(continued)**

(b) In a second experiment, the student uses a motion sensor and computer to produce the following velocity-time graph for the trolley

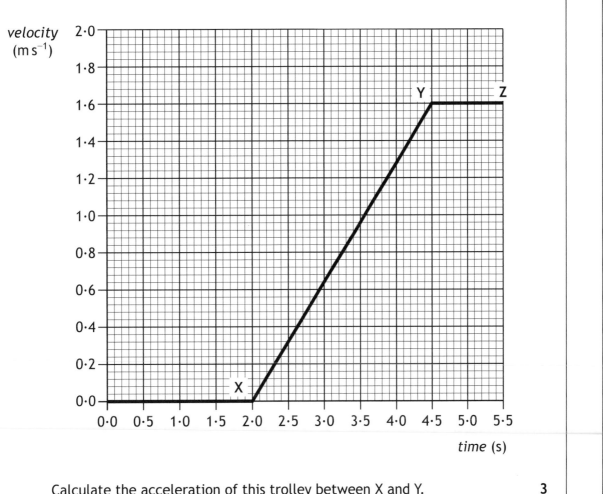

Calculate the acceleration of this trolley between X and Y.   3

Space for working and answer

[Turn over

MARKS | DO NOT WRITE IN THIS MARGIN

9. A child throws a stone horizontally from a bridge into a river.

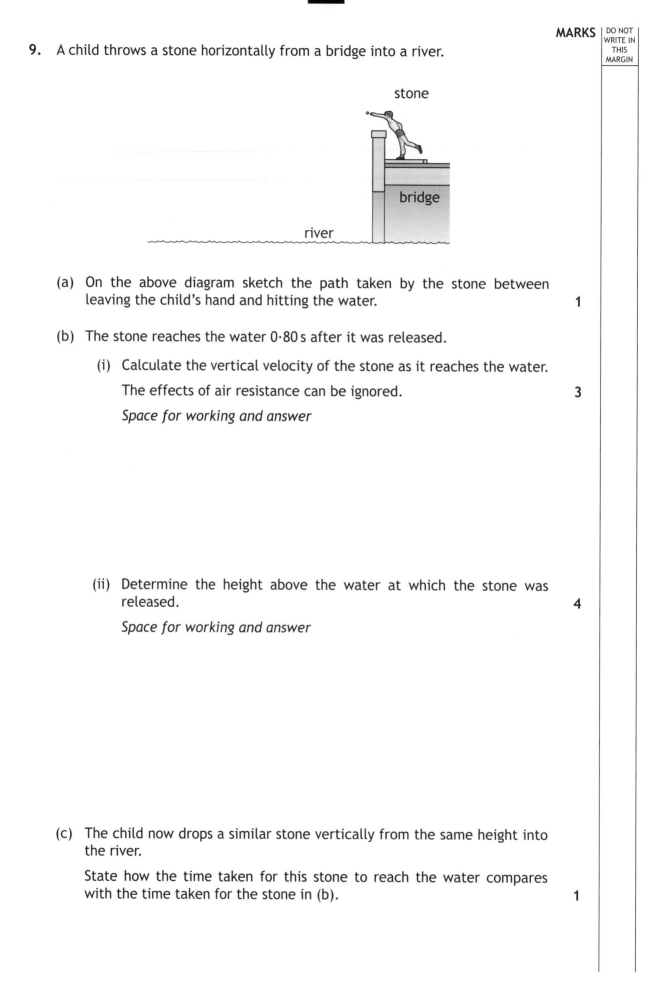

(a) On the above diagram sketch the path taken by the stone between leaving the child's hand and hitting the water.   **1**

(b) The stone reaches the water 0·80 s after it was released.

    (i) Calculate the vertical velocity of the stone as it reaches the water. The effects of air resistance can be ignored.   **3**

    Space for working and answer

    (ii) Determine the height above the water at which the stone was released.   **4**

    Space for working and answer

(c) The child now drops a similar stone vertically from the same height into the river.

State how the time taken for this stone to reach the water compares with the time taken for the stone in (b).   **1**

MARKS | DO NOT WRITE IN THIS MARGIN

10. Space exploration involves placing astronauts in difficult environments. Despite this, many people believe the benefits of space exploration outweigh the risks.

Using your knowledge of physics, comment on the benefits and/or risks of space exploration.

3

[Turn over

[BLANK PAGE]

DO NOT WRITE ON THIS PAGE

MARKS | DO NOT WRITE IN THIS MARGIN

11. Craters on the Moon are caused by meteors striking its surface.

A student investigates how a crater is formed by dropping a marble into a tray of sand.

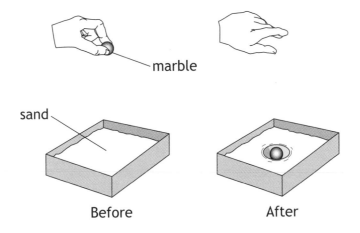

Before   After

(a) The marble has a mass of 0·040 kg.

(i) Calculate the loss in potential energy of the marble when it is dropped from a height of 0·50 m.   **3**

Space for working and answer

(ii) Describe the energy change that takes place as the marble hits the sand.   **1**

[Turn over

MARKS | DO NOT WRITE IN THIS MARGIN

11.   (continued)

(b)   The student drops the marble from different heights and measures the diameter of each crater that is formed.

The table shows the student's results.

height (m)	diameter (m)
0·05	0·030
0·10	0·044
0·15	0·053
0·35	0·074
0·40	0·076
0·45	0·076

(i)   Using the graph paper below, draw a graph of these results.   3

(Additional graph paper, if required, can be found on *Page twenty-eight*)

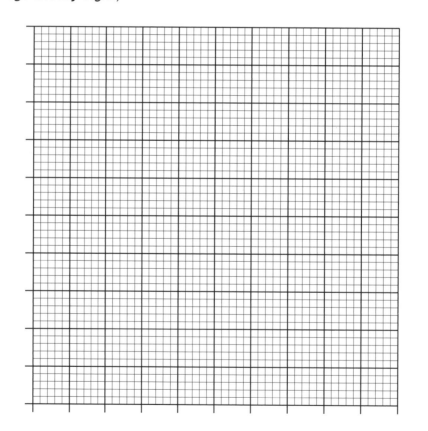

MARKS | DO NOT WRITE IN THIS MARGIN

11. **(b)** **(continued)**

(ii) Use your graph to predict the diameter of the crater that is formed when the marble is dropped from a height of 0·25 m.

1

(iii) Suggest two improvements that the student could make to this investigation.

2

(c) (i) Suggest another variable, which could be investigated, that may affect the diameter of a crater.

1

(ii) Describe experimental work that could be carried out to investigate how this variable affects the diameter of a crater.

2

[END OF QUESTION PAPER]

ADDITIONAL SPACE FOR ANSWERS AND ROUGH WORKING

Additional graph paper for Q11 (b) (i)

MARKS | DO NOT WRITE IN THIS MARGIN

ADDITIONAL SPACE FOR ANSWERS AND ROUGH WORKING

ADDITIONAL SPACE FOR ANSWERS AND ROUGH WORKING

MARKS | DO NOT WRITE IN THIS MARGIN

Page thirty

NATIONAL 5

2016

National Qualifications 2016

X757/75/02

Physics
Section 1 — Questions

TUESDAY, 24 MAY

1:00 PM — 3:00 PM

Instructions for the completion of Section 1 are given on *Page two* of your question and answer booklet X757/75/01.

Record your answers on the answer grid on *Page three* of your question and answer booklet

Reference may be made to the Data Sheet on *Page two* of this booklet and to the Relationships Sheet X757/75/11.

Before leaving the examination room you must give your question and answer booklet to the Invigilator; if you do not, you may lose all the marks for this paper.

DATA SHEET

Speed of light in materials

Material	Speed in $m\,s^{-1}$
Air	$3{\cdot}0 \times 10^8$
Carbon dioxide	$3{\cdot}0 \times 10^8$
Diamond	$1{\cdot}2 \times 10^8$
Glass	$2{\cdot}0 \times 10^8$
Glycerol	$2{\cdot}1 \times 10^8$
Water	$2{\cdot}3 \times 10^8$

Gravitational field strengths

	Gravitational field strength on the surface in $N\,kg^{-1}$
Earth	9·8
Jupiter	23
Mars	3·7
Mercury	3·7
Moon	1·6
Neptune	11
Saturn	9·0
Sun	270
Uranus	8·7
Venus	8·9

Specific latent heat of fusion of materials

Material	Specific latent heat of fusion in $J\,kg^{-1}$
Alcohol	$0{\cdot}99 \times 10^5$
Aluminium	$3{\cdot}95 \times 10^5$
Carbon Dioxide	$1{\cdot}80 \times 10^5$
Copper	$2{\cdot}05 \times 10^5$
Iron	$2{\cdot}67 \times 10^5$
Lead	$0{\cdot}25 \times 10^5$
Water	$3{\cdot}34 \times 10^5$

Specific latent heat of vaporisation of materials

Material	Specific latent heat of vaporisation in $J\,kg^{-1}$
Alcohol	$11{\cdot}2 \times 10^5$
Carbon Dioxide	$3{\cdot}77 \times 10^5$
Glycerol	$8{\cdot}30 \times 10^5$
Turpentine	$2{\cdot}90 \times 10^5$
Water	$22{\cdot}6 \times 10^5$

Speed of sound in materials

Material	Speed in $m\,s^{-1}$
Aluminium	5200
Air	340
Bone	4100
Carbon dioxide	270
Glycerol	1900
Muscle	1600
Steel	5200
Tissue	1500
Water	1500

Specific heat capacity of materials

Material	Specific heat capacity in $J\,kg^{-1}\,°C^{-1}$
Alcohol	2350
Aluminium	902
Copper	386
Glass	500
Ice	2100
Iron	480
Lead	128
Oil	2130
Water	4180

Melting and boiling points of materials

Material	Melting point in °C	Boiling point in °C
Alcohol	−98	65
Aluminium	660	2470
Copper	1077	2567
Glycerol	18	290
Lead	328	1737
Iron	1537	2737

Radiation weighting factors

Type of radiation	Radiation weighting factor
alpha	20
beta	1
fast neutrons	10
gamma	1
slow neutrons	3
X-rays	1

SECTION 1

Attempt ALL questions

1. The symbol for an electronic component is shown.

    This is the symbol for

    A   an LDR

    B   a transistor

    C   an LED

    D   a photovoltaic cell

    E   a thermistor.

2. A uniform electric field exists between plates Q and R.

    The diagram shows the path taken by a particle as it passes through the field.

    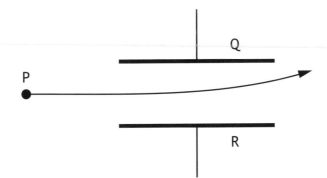

    Which row in the table identifies the charge on the particle, the charge on plate Q and the charge on plate R?

	Charge on particle	Charge on plate Q	Charge on plate R
A	negative	positive	negative
B	negative	negative	positive
C	no charge	negative	positive
D	no charge	positive	negative
E	positive	positive	negative

[Turn over

3.   A circuit is set up as shown.

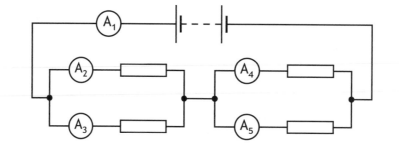

The reading on ammeter A_1 is 5·0A.

The reading on ammeter A_2 is 2·0A.

The reading on ammeter A_4 is 1·0A.

Which row in the table shows the reading on ammeters A_3 and A_5?

	Reading on ammeter A_3 (A)	Reading on ammeter A_5 (A)
A	2·0	1·0
B	3·0	1·0
C	2·0	4·0
D	3·0	4·0
E	5·0	5·0

4.   Two resistors are connected as shown.

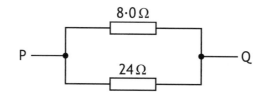

The total resistance between P and Q is

A   0·17 Ω

B   3·0 Ω

C   6·0 Ω

D   16 Ω

E   32 Ω.

5. A block has the dimensions shown.

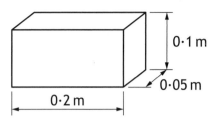

0·1 m

0·05 m

0·2 m

The block is placed so that one of the surfaces is in contact with a smooth table top.

The weight of the block is 4·90 N.

The **minimum** pressure exerted by the block on the table top is

A   25 Pa

B   245 Pa

C   490 Pa

D   980 Pa

E   4900 Pa.

6. A syringe is connected to a pressure meter as shown.

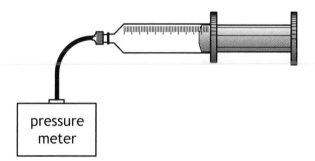

pressure
meter

The syringe contains a fixed mass of air of volume 150 mm³.

The reading on the pressure meter is 120 kPa.

The volume of air inside the syringe is now changed to 100 mm³.

The temperature of the air in the syringe remains constant.

The reading on the pressure meter is now

A   80 kPa

B   125 kPa

C   180 kPa

D   80 000 kPa

E   180 000 kPa.

7. A sample of an ideal gas is enclosed in a sealed container.

Which graph shows how the pressure p of the gas varies with the temperature T of the gas?

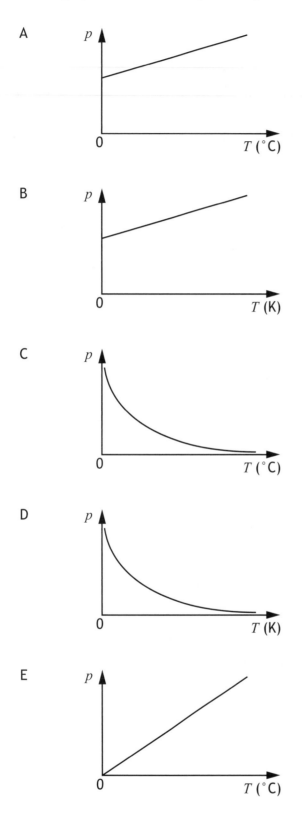

8. A student makes the following statements about waves.

   I   Waves transfer energy.

   II   A wave with a short wavelength diffracts more than a wave with a long wavelength.

   III The amplitude of a wave depends on its wavelength.

   Which of these statements is/are correct?

   A   I only

   B   II only

   C   III only

   D   I and II only

   E   I and III only

9. The diagram represents a wave.

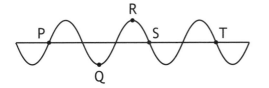

   The wavelength of the wave is the horizontal distance between points

   A   P and Q

   B   P and S

   C   Q and R

   D   R and S

   E   S and T.

[Turn over

10. The diagram represents the position of the crests of waves 3 seconds after a stone is thrown into a pool of still water.

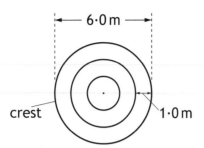

Which row in the table shows the speed and the frequency of the waves?

	Speed (m s^{-1})	Frequency (Hz)
A	0·33	3
B	0·33	1
C	1·0	1
D	1·0	3
E	1·0	4

11. A ray of red light passes through a double glazed window.

Which diagram shows the path of the ray as it passes through the window?

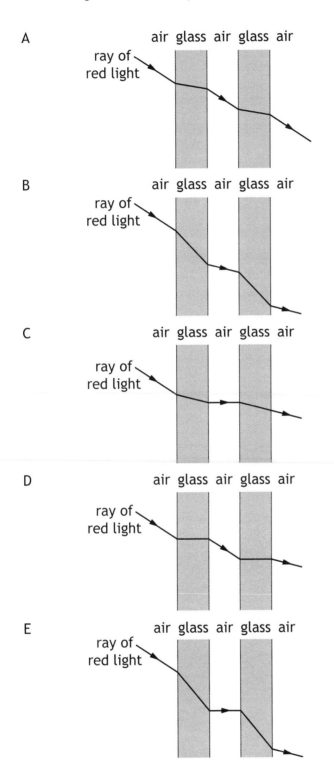

A

air glass air glass air

ray of red light

B

air glass air glass air

ray of red light

C

air glass air glass air

ray of red light

D

air glass air glass air

ray of red light

E

air glass air glass air

ray of red light

[Turn over

12. Which row in the table shows how the mass and charge of an alpha particle compares to the mass and charge of a beta particle?

	Mass of an alpha particle compared to mass of a beta particle	Charge on an alpha particle compared to charge on a beta particle
A	larger	same
B	larger	opposite
C	same	same
D	smaller	opposite
E	smaller	same

13. During ionisation an atom becomes a positive ion.

Which of the following has been removed from the atom?

A   An alpha particle

B   An electron

C   A gamma ray

D   A neutron

E   A proton

14. Which of the following is a vector quantity?

A   Mass

B   Time

C   Speed

D   Kinetic energy

E   Acceleration

15. A ball moves along a horizontal frictionless surface and down a slope as shown.

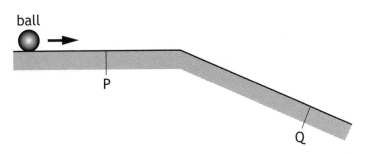

Which of the following graphs shows how the speed of the ball varies with time as it travels from P to Q?

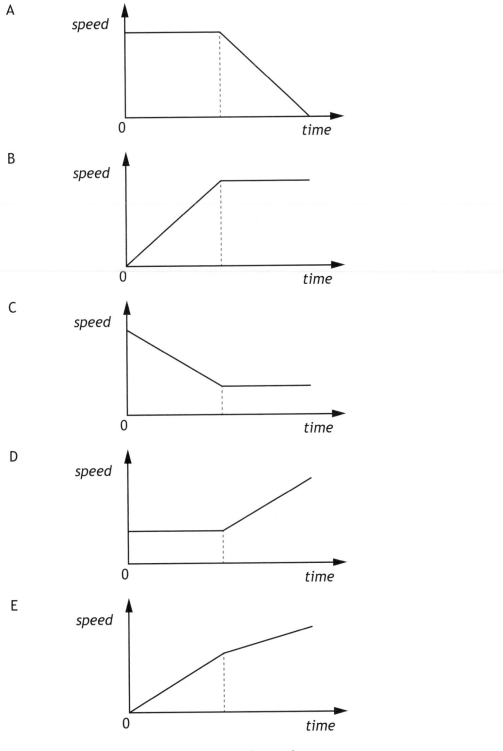

[Turn over

16. A cyclist is travelling at $10\,m\,s^{-1}$ along a level road.

   The cyclist applies the brakes and comes to rest in a time of 5 s.

   The combined mass of the cycle and cyclist is 80 kg.

   The maximum energy converted to heat by the brakes is

   A   160 J

   B   400 J

   C   800 J

   D   4000 J

   E   8000 J.

17. A rocket is taking off from the surface of the Earth. The rocket engines exert a force on the exhaust gases.

   Which of the following is the reaction to this force?

   A   The force of the Earth on the exhaust gases.

   B   The force of the Earth on the rocket engines.

   C   The force of the rocket engines on the Earth.

   D   The force of the exhaust gases on the Earth.

   E   The force of the exhaust gases on the rocket engines.

18. A ball is projected horizontally with a velocity of $1.5\,m\,s^{-1}$ from a cliff as shown.

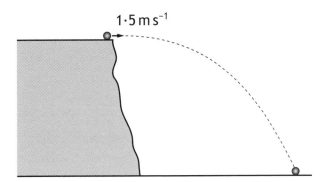

The ball hits the ground $1.2\,s$ after it leaves the cliff.

The effects of air resistance are negligible.

Which row in the table shows the horizontal velocity and vertical velocity of the ball just before it hits the ground?

	Horizontal velocity $(m\,s^{-1})$	Vertical velocity $(m\,s^{-1})$
A	12	12
B	12	1.5
C	1.5	12
D	1.5	13
E	0	12

19. The minimum amount of energy required to change $0.5\,kg$ of water at its boiling point into steam at the same temperature is

A $2.09 \times 10^3\,J$

B $1.67 \times 10^5\,J$

C $3.34 \times 10^5\,J$

D $1.13 \times 10^6\,J$

E $2.26 \times 10^6\,J$.

20. A student makes the following statements about the Universe.

    I   The Big Bang Theory is a theory about the origin of the Universe.

    II   The Universe is approximately 14 million years old.

    III The Universe is expanding.

    Which of these statements is/are correct?

    A   I only
    B   II only
    C   I and II only
    D   I and III only
    E   I, II and III.

**[END OF SECTION 1. NOW ATTEMPT THE QUESTIONS IN SECTION 2 OF
YOUR QUESTION AND ANSWER BOOKLET]**

National
Qualifications
2016

X757/75/11

Physics
Relationships Sheet

TUESDAY, 24 MAY
1:00 PM – 3:00 PM

$$E_p = mgh$$

$$E_k = \frac{1}{2}mv^2$$

$$Q = It$$

$$V = IR$$

$$R_T = R_1 + R_2 + \ldots$$

$$\frac{1}{R_T} = \frac{1}{R_1} + \frac{1}{R_2} + \ldots$$

$$V_2 = \left(\frac{R_2}{R_1 + R_2}\right)V_s$$

$$\frac{V_1}{V_2} = \frac{R_1}{R_2}$$

$$P = \frac{E}{t}$$

$$P = IV$$

$$P = I^2 R$$

$$P = \frac{V^2}{R}$$

$$E_h = cm\Delta T$$

$$p = \frac{F}{A}$$

$$\frac{pV}{T} = \text{constant}$$

$$p_1 V_1 = p_2 V_2$$

$$\frac{p_1}{T_1} = \frac{p_2}{T_2}$$

$$\frac{V_1}{T_1} = \frac{V_2}{T_2}$$

$$d = vt$$

$$v = f\lambda$$

$$T = \frac{1}{f}$$

$$A = \frac{N}{t}$$

$$D = \frac{E}{m}$$

$$H = Dw_R$$

$$\dot{H} = \frac{H}{t}$$

$$s = vt$$

$$d = \bar{v}t$$

$$s = \bar{v}t$$

$$a = \frac{v - u}{t}$$

$$W = mg$$

$$F = ma$$

$$E_w = Fd$$

$$E_h = ml$$

Additional Relationships

Circle

circumference $= 2\pi r$

area $= \pi r^2$

Sphere

area $= 4\pi r^2$

volume $= \frac{4}{3}\pi r^3$

Trigonometry

$\sin\Theta = \dfrac{\text{opposite}}{\text{hypotenuse}}$

$\cos\Theta = \dfrac{\text{adjacent}}{\text{hypotenuse}}$

$\tan\Theta = \dfrac{\text{opposite}}{\text{adjacent}}$

$\sin^2\Theta + \cos^2\Theta = 1$

Electron Arrangements of Elements

Key

Atomic number
Symbol
Electron arrangement
Name

Transition Elements

Group 1 (1)	Group 2 (2)	(3)	(4)	(5)	(6)	(7)	(8)	(9)	(10)	(11)	(12)	Group 3 (13)	Group 4 (14)	Group 5 (15)	Group 6 (16)	Group 7 (17)	Group 0 (18)
1 **H** 1 Hydrogen																	2 **He** 2 Helium
3 **Li** 2,1 Lithium	4 **Be** 2,2 Beryllium											5 **B** 2,3 Boron	6 **C** 2,4 Carbon	7 **N** 2,5 Nitrogen	8 **O** 2,6 Oxygen	9 **F** 2,7 Fluorine	10 **Ne** 2,8 Neon
11 **Na** 2,8,1 Sodium	12 **Mg** 2,8,2 Magnesium											13 **Al** 2,8,3 Aluminium	14 **Si** 2,8,4 Silicon	15 **P** 2,8,5 Phosphorus	16 **S** 2,8,6 Sulfur	17 **Cl** 2,8,7 Chlorine	18 **Ar** 2,8,8 Argon
19 **K** 2,8,8,1 Potassium	20 **Ca** 2,8,8,2 Calcium	21 **Sc** 2,8,9,2 Scandium	22 **Ti** 2,8,10,2 Titanium	23 **V** 2,8,11,2 Vanadium	24 **Cr** 2,8,13,1 Chromium	25 **Mn** 2,8,13,2 Manganese	26 **Fe** 2,8,14,2 Iron	27 **Co** 2,8,15,2 Cobalt	28 **Ni** 2,8,16,2 Nickel	29 **Cu** 2,8,18,1 Copper	30 **Zn** 2,8,18,2 Zinc	31 **Ga** 2,8,18,3 Gallium	32 **Ge** 2,8,18,4 Germanium	33 **As** 2,8,18,5 Arsenic	34 **Se** 2,8,18,6 Selenium	35 **Br** 2,8,18,7 Bromine	36 **Kr** 2,8,18,8 Krypton
37 **Rb** 2,8,18,8,1 Rubidium	38 **Sr** 2,8,18,8,2 Strontium	39 **Y** 2,8,18,9,2 Yttrium	40 **Zr** 2,8,18,10,2 Zirconium	41 **Nb** 2,8,18,12,1 Niobium	42 **Mo** 2,8,18,13,1 Molybdenum	43 **Tc** 2,8,18,13,2 Technetium	44 **Ru** 2,8,18,15,1 Ruthenium	45 **Rh** 2,8,18,16,1 Rhodium	46 **Pd** 2,8,18,18,0 Palladium	47 **Ag** 2,8,18,18,1 Silver	48 **Cd** 2,8,18,18,2 Cadmium	49 **In** 2,8,18,18,3 Indium	50 **Sn** 2,8,18,18,4 Tin	51 **Sb** 2,8,18,18,5 Antimony	52 **Te** 2,8,18,18,6 Tellurium	53 **I** 2,8,18,18,7 Iodine	54 **Xe** 2,8,18,18,8 Xenon
55 **Cs** 2,8,18,18,8,1 Caesium	56 **Ba** 2,8,18,18,8,2 Barium	57 **La** 2,8,18,18,9,2 Lanthanum	72 **Hf** 2,8,18,32,10,2 Hafnium	73 **Ta** 2,8,18,32,11,2 Tantalum	74 **W** 2,8,18,32,12,2 Tungsten	75 **Re** 2,8,18,32,13,2 Rhenium	76 **Os** 2,8,18,32,14,2 Osmium	77 **Ir** 2,8,18,32,15,2 Iridium	78 **Pt** 2,8,18,32,17,1 Platinum	79 **Au** 2,8,18,32,18,1 Gold	80 **Hg** 2,8,18,32,18,2 Mercury	81 **Tl** 2,8,18,32,18,3 Thallium	82 **Pb** 2,8,18,32,18,4 Lead	83 **Bi** 2,8,18,32,18,5 Bismuth	84 **Po** 2,8,18,32,18,6 Polonium	85 **At** 2,8,18,32,18,7 Astatine	86 **Rn** 2,8,18,32,18,8 Radon
87 **Fr** 2,8,18,32,18,8,1 Francium	88 **Ra** 2,8,18,32,18,8,2 Radium	89 **Ac** 2,8,18,32,18,9,2 Actinium	104 **Rf** 2,8,18,32,32,10,2 Rutherfordium	105 **Db** 2,8,18,32,32,11,2 Dubnium	106 **Sg** 2,8,18,32,32,12,2 Seaborgium	107 **Bh** 2,8,18,32,32,13,2 Bohrium	108 **Hs** 2,8,18,32,32,14,2 Hassium	109 **Mt** 2,8,18,32,32,15,2 Meitnerium	110 **Ds** 2,8,18,32,32,17,1 Darmstadtium	111 **Rg** 2,8,18,32,32,18,1 Roentgenium	112 **Cn** 2,8,18,32,32,18,2 Copernicium						

Lanthanides

57 **La** 2,8,18,18,9,2 Lanthanum	58 **Ce** 2,8,18,20,8,2 Cerium	59 **Pr** 2,8,18,21,8,2 Praseodymium	60 **Nd** 2,8,18,22,8,2 Neodymium	61 **Pm** 2,8,18,23,8,2 Promethium	62 **Sm** 2,8,18,24,8,2 Samarium	63 **Eu** 2,8,18,25,8,2 Europium	64 **Gd** 2,8,18,25,9,2 Gadolinium	65 **Tb** 2,8,18,27,8,2 Terbium	66 **Dy** 2,8,18,28,8,2 Dysprosium	67 **Ho** 2,8,18,29,8,2 Holmium	68 **Er** 2,8,18,30,8,2 Erbium	69 **Tm** 2,8,18,31,8,2 Thulium	70 **Yb** 2,8,18,32,8,2 Ytterbium	71 **Lu** 2,8,18,32,9,2 Lutetium

Actinides

89 **Ac** 2,8,18,32,18,9,2 Actinium	90 **Th** 2,8,18,32,18,10,2 Thorium	91 **Pa** 2,8,18,32,20,9,2 Protactinium	92 **U** 2,8,18,32,21,9,2 Uranium	93 **Np** 2,8,18,32,22,9,2 Neptunium	94 **Pu** 2,8,18,32,24,8,2 Plutonium	95 **Am** 2,8,18,32,25,8,2 Americium	96 **Cm** 2,8,18,32,25,9,2 Curium	97 **Bk** 2,8,18,32,27,8,2 Berkelium	98 **Cf** 2,8,18,32,28,8,2 Californium	99 **Es** 2,8,18,32,29,8,2 Einsteinium	100 **Fm** 2,8,18,32,30,8,2 Fermium	101 **Md** 2,8,18,32,31,8,2 Mendelevium	102 **No** 2,8,18,32,32,8,2 Nobelium	103 **Lr** 2,8,18,32,32,9,2 Lawrencium

N5

Mark

National
Qualifications
2016

X757/75/01

**Physics
Section 1 — Answer Grid
and Section 2**

TUESDAY, 24 MAY

1:00 PM — 3:00 PM

Fill in these boxes and read what is printed below.

Full name of centre

Town

Forename(s)

Surname

Number of seat

Date of birth

Day	Month	Year

Scottish candidate number

Total marks — 110

SECTION 1 — 20 marks
Attempt ALL questions.
Instructions for completion of Section 1 are given on *Page two*.

SECTION 2 — 90 marks
Attempt ALL questions.

Reference may be made to the Data Sheet on *Page two* of the question paper X757/75/02 and to the Relationships Sheet X757/75/11.

Write your answers clearly in the spaces provided in this booklet. Additional space for answers and rough work is provided at the end of this booklet. If you use this space you must clearly identify the question number you are attempting. Any rough work must be written in this booklet. You should score through your rough work when you have written your final copy.

Use **blue** or **black** ink.

Before leaving the examination room you must give this booklet to the Invigilator; if you do not, you may lose all the marks for this paper.

SECTION 1 — 20 marks

The questions for Section 1 are contained in the question paper X757/75/02.

Read these and record your answers on the answer grid on *Page three* opposite.

Use **blue** or **black** ink. Do NOT use gel pens or pencil.

1. The answer to each question is **either** A, B, C, D or E. Decide what your answer is, then fill in the appropriate bubble (see sample question below).

2. There is **only one correct** answer to each question.

3. Any rough work must be written in the additional space for answers and rough work at the end of this booklet.

Sample Question

The energy unit measured by the electricity meter in your home is the:

    A   ampere

    B   kilowatt-hour

    C   watt

    D   coulomb

    E   volt.

The correct answer is **B** — kilowatt-hour. The answer **B** bubble has been clearly filled in (see below).

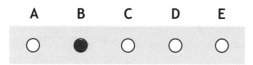

Changing an answer

If you decide to change your answer, cancel your first answer by putting a cross through it (see below) and fill in the answer you want. The answer below has been changed to **D**.

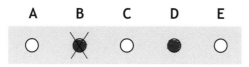

If you then decide to change back to an answer you have already scored out, put a tick (✓) to the **right** of the answer you want, as shown below:

                               or

SECTION 1 — Answer Grid

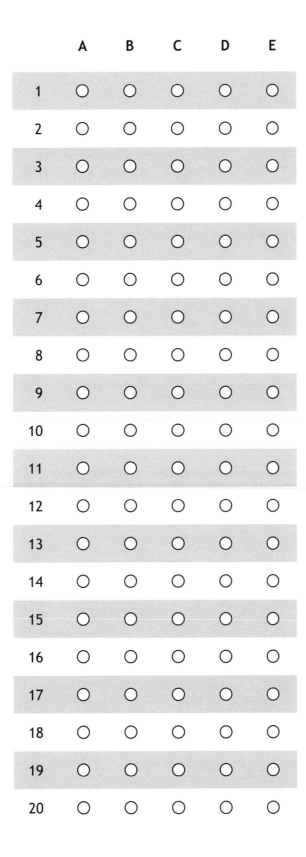

[Turn over

[BLANK PAGE]

DO NOT WRITE ON THIS PAGE

[TURN OVER FOR NEXT QUESTION]

DO NOT WRITE ON THIS PAGE

MARKS | DO NOT WRITE IN THIS MARGIN

SECTION 2 — 90 marks

Attempt ALL questions

1. Electrical storms occur throughout the world.

During one lightning strike 24 C of charge is transferred to the ground in 0·0012 s.

(a) Calculate the average current during the lightning strike. **3**

Space for working and answer

(b) The charge on an electron is $-1·6 \times 10^{-19}$ C.

Determine the number of electrons transferred during the lightning strike. **1**

Space for working and answer

MARKS | DO NOT WRITE IN THIS MARGIN

1. (continued)

    (c) Many tall buildings have a thick strip of metal attached to the side of the building.

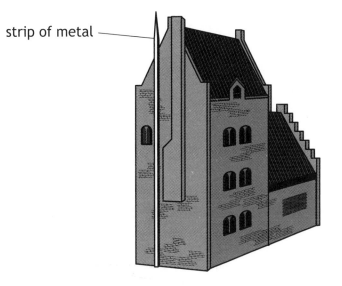

    strip of metal

    This strip is used to protect the building from damage during electrical storms.

    Explain how this strip protects the building from damage.   2

[Turn over

MARKS | DO NOT WRITE IN THIS MARGIN

2. A student investigates the resistance of a resistor using the circuit shown.

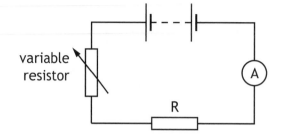

(a) Complete the circuit diagram to show where a voltmeter must be connected to measure the voltage across resistor R.   1

(An additional diagram, if required, can be found on *Page thirty-three*.)

(b) Describe how the student obtains a range of values of voltage and current.   1

MARKS | DO NOT WRITE IN THIS MARGIN

2.   (continued)

(c)   The results of the student's investigation are shown.

Voltage across resistor R (V)	Current in resistor R (A)
1·0	0·20
2·5	0·50
3·2	0·64
6·2	1·24

Use **all** these results to determine the resistance of resistor R.   4

Space for working and answer

(d)   The student now replaces resistor R with a filament lamp and repeats the investigation. A sketch graph of the student's results is shown.

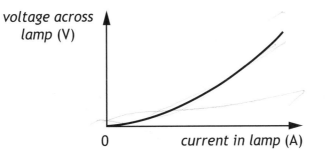

State a conclusion that can be made about the resistance of the filament lamp.   1

[Turn over

3. A washing machine fills with water at a temperature of 15·0 °C.

The water is heated by a heating element.

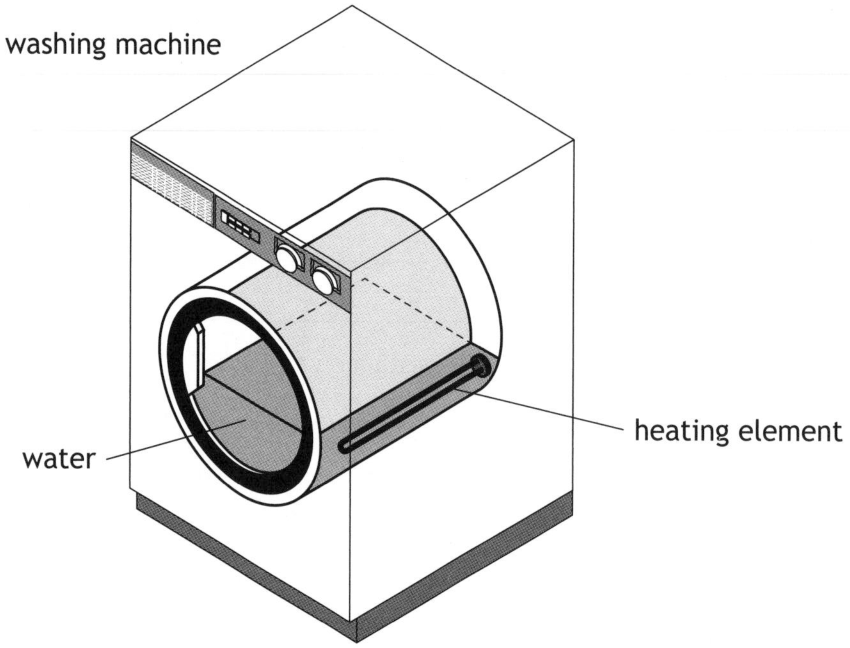

washing machine

heating element

water

(a) The mass of the water in the washing machine is 6·00 kg.

Show that the minimum energy required to increase the temperature of the water from 15·0 °C to 40·0 °C is 627 000 J.

2

Space for working and answer

MARKS | DO NOT WRITE IN THIS MARGIN

3. (continued)

(b) The heating element has a power rating of 1800 W.

(i) Calculate the time taken for the heating element to supply the energy calculated in (a).

3

Space for working and answer

(ii) Explain why, in practice, it takes longer to heat the water from 15 °C to 40 °C than calculated in (b)(i).

1

[Turn over

MARKS | DO NOT WRITE IN THIS MARGIN

3. **(continued)**

(c) The temperature of the water in the washing machine is monitored by a circuit containing a thermistor.

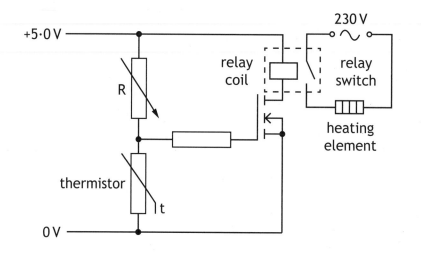

As the temperature of the water increases, the resistance of the thermistor decreases.

The heating element is switched off when the temperature of the water reaches 40 °C.

Explain how the circuit operates to **switch off** the heating element.   **3**

MARKS | DO NOT WRITE IN THIS MARGIN

4. The diagram shows some parts of the electromagnetic spectrum in order of increasing wavelength.

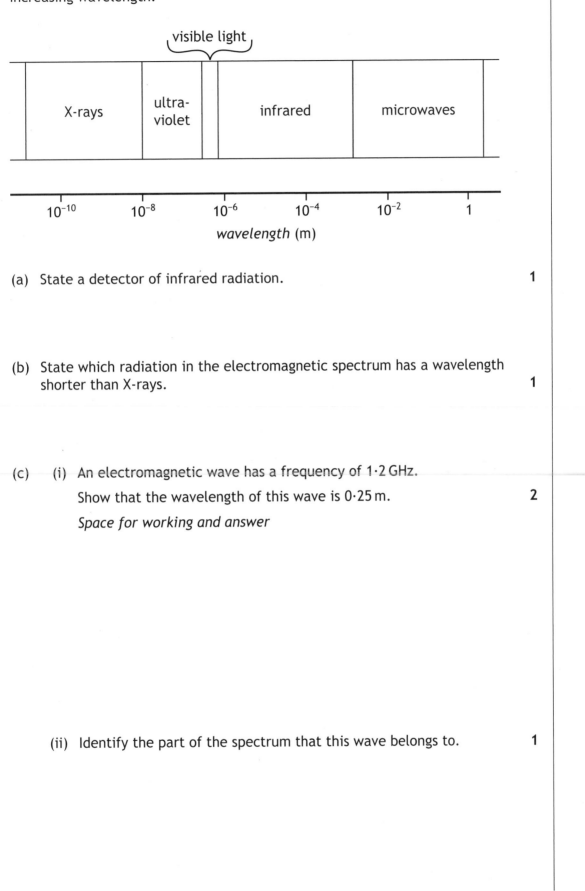

(a) State a detector of infrared radiation.

1

(b) State which radiation in the electromagnetic spectrum has a wavelength shorter than X-rays.

1

(c) (i) An electromagnetic wave has a frequency of 1·2 GHz.

Show that the wavelength of this wave is 0·25 m.

Space for working and answer

2

(ii) Identify the part of the spectrum that this wave belongs to.

1

[Turn over

[BLANK PAGE]

DO NOT WRITE ON THIS PAGE

MARKS | DO NOT WRITE IN THIS MARGIN

5. A Physics textbook contains the following statement.

    "Electromagnetic waves can be sent out like ripples on a pond."

    Using your knowledge of physics, comment on the similarities and/or differences between electromagnetic waves and the ripples on a pond.

    3

MARKS | DO NOT WRITE IN THIS MARGIN

6. A student directs a ray of red light into a Perspex block to investigate refraction.

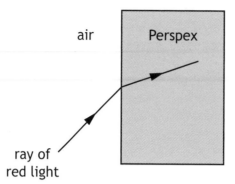

ray of
red light

(a) On the diagram, draw and label:

    (i) the normal;   **1**

    (ii) the angle of incidence i and the angle of refraction r.   **1**

    (An additional diagram, if required, can be found on *Page thirty-three*)

(b) The student varies the angle of incidence and measures the corresponding angles of refraction. The results are plotted on a graph.

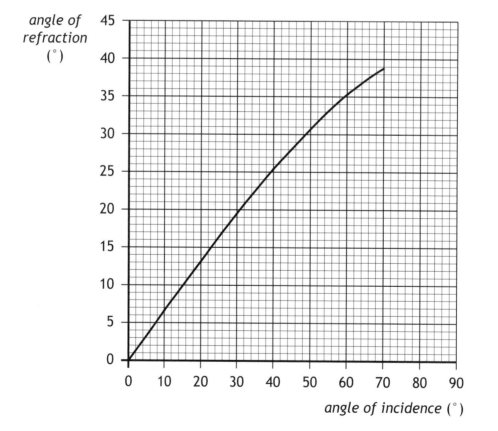

MARKS | DO NOT WRITE IN THIS MARGIN

6. (b) (continued)

(i) Determine the angle of refraction when the angle of incidence is 12°.

1

(ii) Use the graph to predict the angle of refraction the student would obtain for an angle of incidence of 80°.

1

(c) Suggest why it would be good practice for the student to repeat the investigation a further three or four times.

1

Page seventeen

[Turn over

MARKS | DO NOT WRITE IN THIS MARGIN

7. A spacecraft uses a radioisotope thermoelectric generator (RTG) as a power source.

— RTG

The RTG transforms the heat released by the radioactive decay of plutonium-238 into electrical energy.

(a) In 15 minutes, $7 \cdot 92 \times 10^{18}$ nuclei of plutonium-238 decay.

Calculate the activity of the plutonium-238.   **3**

Space for working and answer

(b) Each decay produces heat that is transformed into $4 \cdot 49 \times 10^{-14}$ J of electrical energy.

Determine the power output of the RTG.   **2**

Space for working and answer

MARKS | DO NOT WRITE IN THIS MARGIN

7. **(continued)**

(c) Plutonium-238 emits alpha radiation.

Explain why a source that emits alpha radiation requires less shielding than a source that emits gamma radiation.   1

MARKS | DO NOT WRITE IN THIS MARGIN

8. During medical testing a beta source is used to irradiate a sample of tissue of mass 0·50 kg from a distance of 0·10 m.

The sample absorbs $9·6 \times 10^{-5}$ J of energy from the beta source.

beta source

tissue sample

(a) (i) Calculate the absorbed dose received by the sample.   **3**

Space for working and answer

(ii) Calculate the equivalent dose received by the sample.   **3**

Space for working and answer

MARKS | DO NOT WRITE IN THIS MARGIN

8. (continued)

(b) The beta source used during testing has a half-life of 36 hours.

The initial activity of the beta source is 12 kBq.

Determine the activity of the source 144 hours later.   3

Space for working and answer

[Turn over

9. A student walks around a building from point X to point Y.

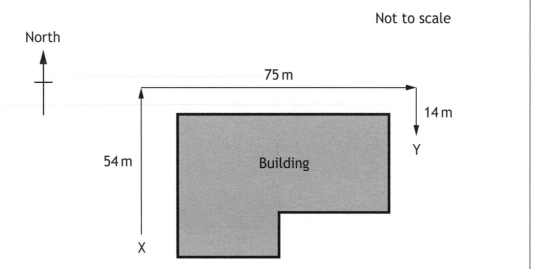

Not to scale

North

75 m

14 m

Y

54 m

Building

X

(a) By scale diagram, or otherwise, determine:

    (i) the magnitude of the displacement of the student from point X to point Y;

    Space for working and answer

2

    (ii) the direction of displacement of the student from point X to point Y.

    Space for working and answer

2

MARKS | DO NOT WRITE IN THIS MARGIN

9. (continued)

(b) The student takes 68 s to travel from point X to point Y.

(i) Determine the average velocity of the student from point X to point Y.

3

Space for working and answer

(ii) The student states that their average speed between point X and point Y is greater than the magnitude of their average velocity between point X and point Y.

Explain why the student is correct.

2

10. An air descender is a machine that controls the rate at which a climber drops from a platform at the top of a climbing wall.

A climber, attached to the air descender by a rope, steps off the platform and drops towards the ground and lands safely.

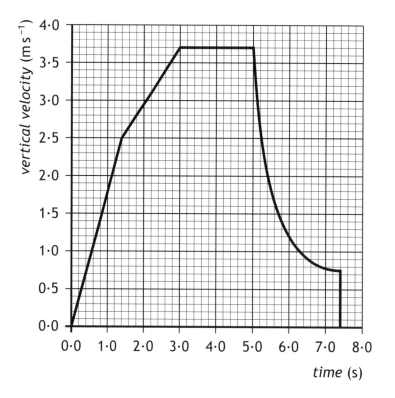

air descender

rope

platform

air descender

not to scale

ground

The graph shows how the vertical velocity of the climber varies with time from the instant the climber leaves the platform until landing.

MARKS | DO NOT WRITE IN THIS MARGIN

10. (continued)

(a) Calculate the acceleration of the climber during the first 1·4 s of the drop.

3

Space for working and answer

(b) Calculate the distance the climber drops during the first 3·0 s.

3

Space for working and answer

(c) During part of the drop the forces on the climber are balanced.

On the diagram below show all the forces acting vertically on the climber during this part of the drop.

You must name these forces **and** show their directions.

3

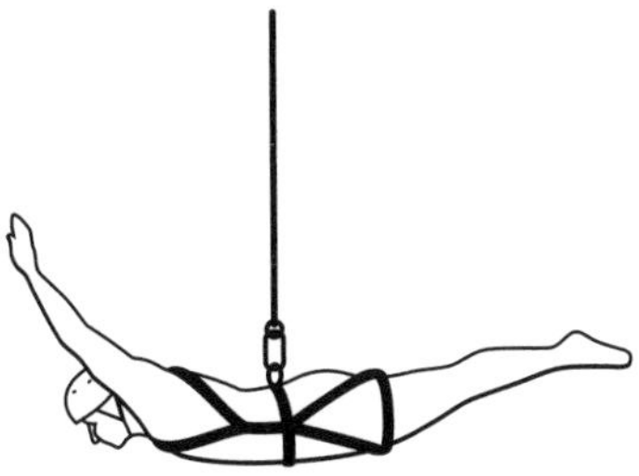

(An additional diagram, if required, can be found on *Page thirty-three*)

Page twenty-six

[BLANK PAGE]

DO NOT WRITE ON THIS PAGE

MARKS | DO NOT WRITE IN THIS MARGIN

11. The length of runway required for aircraft to lift off the ground into the air is known as the ground roll.

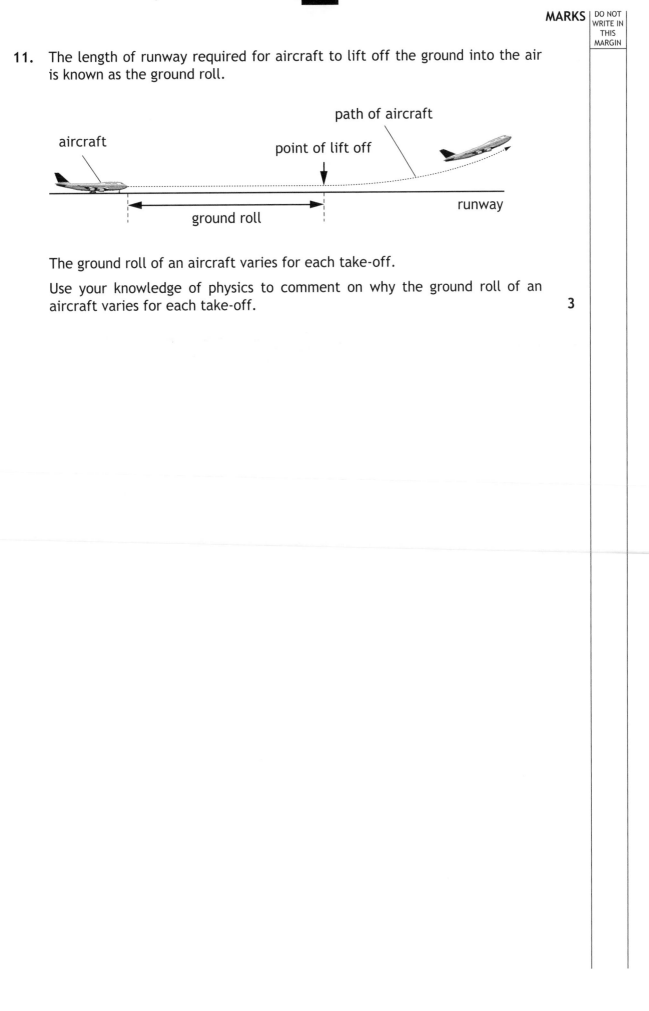

The ground roll of an aircraft varies for each take-off.

Use your knowledge of physics to comment on why the ground roll of an aircraft varies for each take-off.

3

MARKS | DO NOT WRITE IN THIS MARGIN

12. On 12th November 2014, on a mission known as Rosetta, the European Space Agency successfully landed a probe on the surface of a comet.

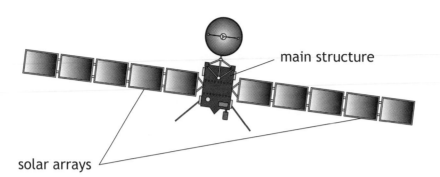

main structure

solar arrays

The main structure of the Rosetta spacecraft consists of an orbiter, a lander and propellant.

Rosetta spacecraft data		
Launch mass	Orbiter Lander Propellant	$1 \cdot 23 \times 10^3$ kg $0 \cdot 10 \times 10^3$ kg $1 \cdot 67 \times 10^3$ kg
	Total	$3 \cdot 00 \times 10^3$ kg
Energy source	Solar array output	850 W at 3·4 AU 395 W at 5·25 AU
Trajectory control	24 Thrusters	10 N of force each

(a) Calculate the total weight of the spacecraft on Earth.   **3**

Space for working and answer

(b) The solar arrays contain photovoltaic cells.

   (i) State the energy change in a photovoltaic cell.   **1**

   (ii) Suggest why the solar arrays were designed so that they can rotate.   **1**

MARKS | DO NOT WRITE IN THIS MARGIN

12. (b) (continued)

(iii) Calculate the total energy output of the solar arrays when operating at 5·25 AU for 2 hours.

Space for working and answer

3

(c) At a point on its journey between Earth and the comet, the spacecraft was travelling at a constant velocity.

(i) The spacecraft switched on four of its thrusters to accelerate it in the direction of travel.

The four thrusters exerted a force on the spacecraft in the same direction.

Determine the total force produced by these thrusters.

Space for working and answer

1

(ii) At this point, the spacecraft had used $1·00 \times 10^3$ kg of propellant.

Calculate the acceleration of the spacecraft.

Space for working and answer

4

[Turn over for next question

MARKS | DO NOT WRITE IN THIS MARGIN

13. Read the passage and answer the questions that follow.

Supernova explosion

The average temperature of the surface of the Sun is 5778 K. In the core of the Sun energy is produced by nuclear fusion. Once the Sun has used all its nuclear fuel it will collapse to form a white dwarf.

A star with a mass much larger than that of the Sun will end its life in an enormous explosion called a supernova. The energy released in a supernova explosion is more than a hundred times the energy that the Sun will radiate over its entire 10 billion year lifetime.

In our galaxy, the star Betelgeuse is predicted to explode in a supernova. Betelgeuse has a mass of around 8 times the mass of the Sun. Even though Betelgeuse is 640 light-years from Earth, the supernova will be as bright as a full moon at night in our sky.

(a) State what is meant by the term *nuclear fusion*. 1

(b) Determine the average temperature of the surface of the Sun in degrees Celsius. 1

Space for working and answer

MARKS | DO NOT WRITE IN THIS MARGIN

13.   (continued)

(c)   Show that the distance from Earth to Betelgeuse is 6.1×10^{18} m.   **3**

Space for working and answer

(d)   Betelgeuse may have already exploded in a supernova.
Explain this statement.   **1**

[END OF QUESTION PAPER]

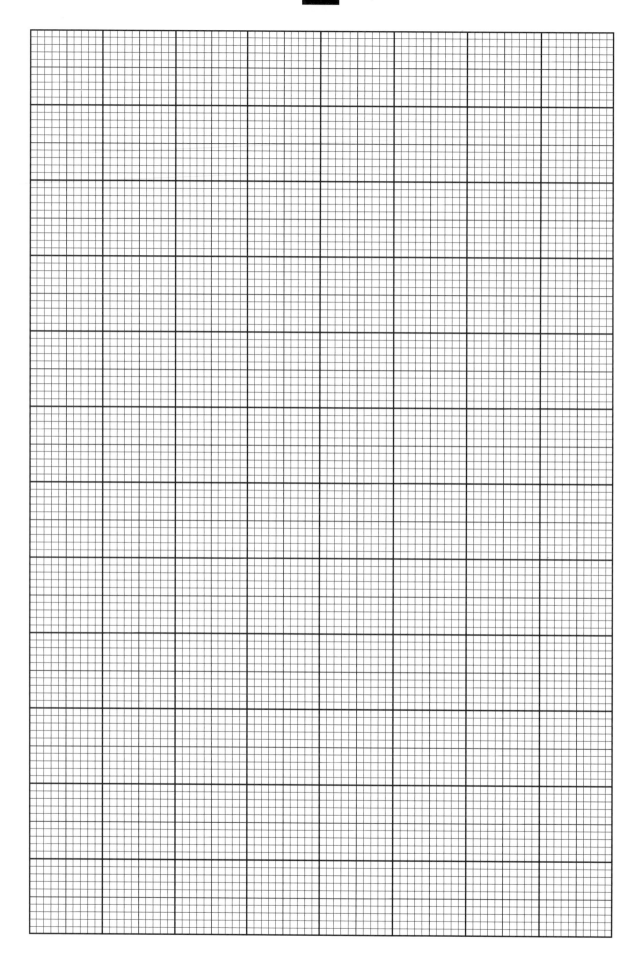

ADDITIONAL SPACE FOR ANSWERS AND ROUGH WORKING

Additional diagram for Q2 (a)

Additional diagram for Q6 (a)

Additional diagram for Q10 (c)

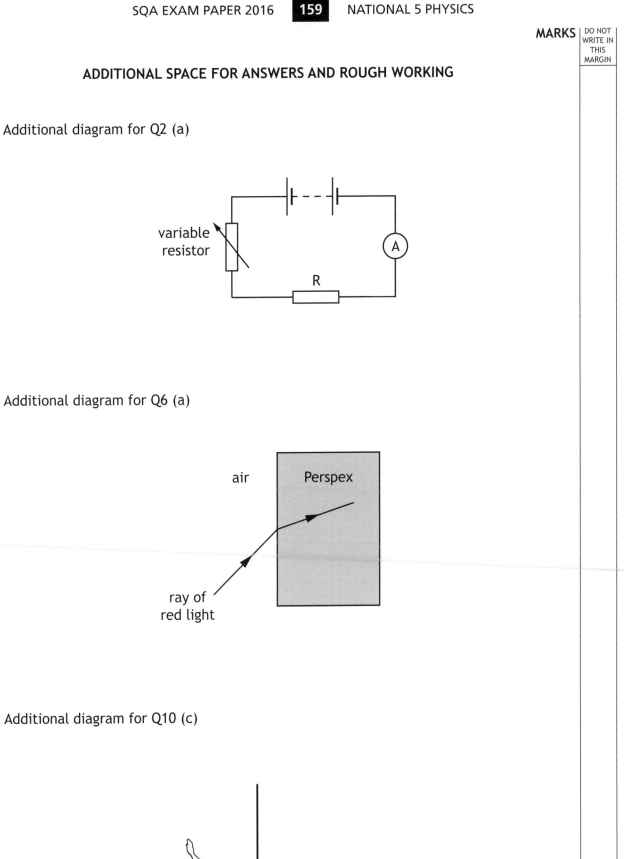

MARKS | DO NOT WRITE IN THIS MARGIN

ADDITIONAL SPACE FOR ANSWERS AND ROUGH WORKING

MARKS | DO NOT WRITE IN THIS MARGIN

ADDITIONAL SPACE FOR ANSWERS AND ROUGH WORKING

[BLANK PAGE]

DO NOT WRITE ON THIS PAGE

NATIONAL 5

Answers

NATIONAL 5 PHYSICS
2014

Section 1

1.	D	6.	A	11.	E	16.	D
2.	D	7.	A	12.	A	17.	D
3.	B	8.	C	13.	E	18.	E
4.	C	9.	B	14.	A	19.	C
5.	B	10.	B	15.	E	20.	D

Section 2

1. (a) $P = \dfrac{V^2}{R}$

$= \dfrac{12 \cdot 0^2}{100}$

$= 1 \cdot 44$ W

(b) (i) $\dfrac{1}{R_T} = \dfrac{1}{R_1} + \dfrac{1}{R_2} + \dfrac{1}{R_3}$

$\dfrac{1}{R_T} = \dfrac{1}{100} + \dfrac{1}{50} + \dfrac{1}{50}$

$\dfrac{1}{R_T} = \dfrac{1}{20}$

$R_T = 20 \ \Omega$

(ii) **Effect:**

The other lamp:
- remains lit
- stays on
- is the same brightness
- gets brighter
- is not affected

Justification:

The current still has a path through the other lamp.

OR

The current in the other lamp is the same (only acceptable if other lamp stays same brightness).

OR

The current in the other lamp is greater (only acceptable if other lamp gets brighter).

OR

It has the same voltage/12 V (across it).

OR

The lamps are connected in parallel.

2. (a) (i) $V_2 = V_S - V_1 = 3 \cdot 0$ (V)

$I = \dfrac{V_2}{R}$

$= \dfrac{3 \cdot 0}{1050}$

$= (2 \cdot 857 \times 10^{-3}$ A$)$

$R_1 = \dfrac{V_1}{I}$

$= \dfrac{2 \cdot 0}{2 \cdot 857 \times 10^{-3}}$

$= 700 \ \Omega$

(ii) 80 °C

(b) (i) As R_{th} increases, V_{th} increases
MOSFET/transistor turns on
Relay switches on (the heater).

(ii) Temperature decreases
Resistance of thermistor must be greater/increase to switch on MOSFET/transistor.

3. (a) $E = Pt$
$E = 15 \times 10 \times 60$
$E = 9000$ J

(b) (i) X

(ii) $E = cm\Delta T$
$9000 = c \times 1 \cdot 0 \times 10$
$c = 900$ J kg^{-1} °C^{-1}

(c) (i) Insulating the metal block
OR
Switch heater on for shorter time

(ii) Increase/greater (for insulating)
OR
Decrease/lower (for shorter time)

4. (a) $f = N° \ of \ waves/time$

$= \dfrac{4}{20}$

$= 0 \cdot 2$ (Hz)

$v = f\lambda$

$= 0 \cdot 2 \times 12$

$= 2 \cdot 4$ m s^{-1}

(b)

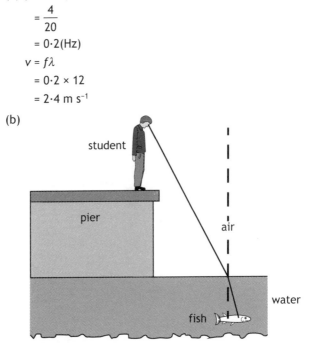

student

pier

air

water

fish

(1) mark for ray changing direction at water/air boundary.
(1) mark for angle in water less than angle in air.
Angle of incidence in water should be less than the angle of refraction in air.
(1) mark for correct normal (must be placed at the point where a ray meets the water/air boundary).

5. (a) *UV index = (total effect of UV radiation x elevation above sea level adjustment x cloud adjustment) ÷ 25*

UV index = (280 × 1·12 × 0·31) ÷ 25
= 3·89
= 4

(b)

	UVA	UVB	UVC
Type of sunscreen that absorbs most of this radiation	**P**	Q	**R**
Type of sunscreen that absorbs least of this radiation	R	**R**	**P**

(c) Detecting counterfeit bank notes, setting dental fillings, etc

6. (a) The time taken for the activity / corrected count rate of a radioactive source to half.

(b) (i) Measure the count in a set time interval
Repeat at regular intervals
Measure background count and subtract

(ii) Half-life = 10 minutes

(iii) 88 → 44 → 22 → 11 → 5·5
mark for evidence of halving
Count rate = 5·5 counts per minute

7. This is an open-ended question.
1 mark: The student has demonstrated a limited understanding of the physics involved. The student has made some statement(s) which is/are relevant to the situation, showing that at least a little of the physics within the problem is understood.
2 marks: The student has demonstrated a reasonable understanding of the physics involved. The student makes some statement(s) which is/are relevant to the situation, showing that the problem is understood.
3 marks: The maximum available mark would be awarded to a student who has demonstrated a good understanding of the physics involved. The student shows a good comprehension of the physics of the situation and has provided a logically correct answer to the question posed. This type of response might include a statement of the principles involved, a relationship or an equation, and the application of these to respond to the problem. This does not mean the answer has to be what might be termed an "excellent" answer or a "complete" one.

8. (a) (i) $D = \dfrac{E}{m}$

$= \dfrac{7·2 \times 10^{-3}}{80·0}$

$= 9·0 \times 10^{-5}$ Gy

(ii) $H = Dw_R$

$= 9·0 \times 10^{-5} \times 1$

$= 9·0 \times 10^{-5}$ Sv

(b) When an **atom** gains or loses electrons.

9. This is an open-ended question.
1 mark: The student has demonstrated a limited understanding of the physics involved. The student has made some statement(s) which is/are relevant to the situation, showing that at least a little of the physics within the problem is understood.
2 marks: The student has demonstrated a reasonable understanding of the physics involved. The student makes some statement(s) which is/are relevant to the situation, showing that the problem is understood.
3 marks: The maximum available mark would be awarded to a student who has demonstrated a good understanding of the physics involved. The student shows a good comprehension of the physics of the situation and has provided a logically correct answer to the question posed. This type of response might include a statement of the principles involved, a relationship or an equation, and the application of these to respond to the problem. This does not mean the answer has to be what might be termed an "excellent" answer or a "complete" one.

10. (a) (i) $a = \dfrac{v - u}{t}$

$= \dfrac{4·8 - 0}{25}$

$= 0·19$ m s^{-2}

(ii) constant speed
OR
constant velocity

(iii)
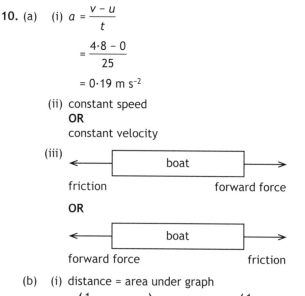
friction ← [boat] → forward force

OR

forward force ← [boat] → friction

(b) (i) distance = area under graph

$= \left(\dfrac{1}{2} \times 25 \times 4·8\right) + (4·8 \times 425) + \left(\dfrac{1}{2} \times 60 \times 4·8\right)$

$(= 60 + 2040 + 144)$

$= 2244$ m

(ii) v = total distance/time
$= 2244/510$
$= 4·4$ m s^{-1}

11. (a) To check that the maximum take-off weight is not exceeded.

(b) 19 625 N

(c) $d = vt$
$201\,000 = 67 \times t$
$t = 3\,000$ s

12. (a) $W = mg$
           $= 0.94 \times 9.8$
           $= 9.2$ N

(b) **Method 1**
    $A = 3 \times (2.0 \times 10^{-4})$
    $= 6.0 \times 10^{-4}$ (m^2)

    $p = \dfrac{F}{A}$

    $= \dfrac{9.2}{6.0 \times 10^{-4}}$

    $= 1.5 \times 10^4$ Pa

OR

Method 2

    $p = \dfrac{F}{A}$

    $= \dfrac{9.2}{2.0 \times 10^{-4}}$

    $= 4.6 \times 10^4$ Pa

(If this line is the candidate's final answer, unit required.)

    total $p = \dfrac{4.6 \times 10^4}{A}$

    $= 1.5 \times 10^4$ Pa

OR

Method 3...
take $\frac{1}{3}$ of weight and use this for
F in $p = F/A$

(c) Rocket/bottle pushes down on water, water pushes up on rocket/bottle.

(d) F_{un} = upthrust – weight
    $= 370 - 9.2$
    $= 360.8$ (N)

    $a = \dfrac{F}{m}$

    $= \dfrac{360.8}{0.94}$

    $= 380$ m s^{-2}

(e) • more water will increase weight/mass
    • unbalanced force decreases
    • acceleration is less

NATIONAL 5 PHYSICS 2015

Section 1

1.	A	6.	D	11.	E	16.	C
2.	A	7.	D	12.	A	17.	A
3.	C	8.	A	13.	E	18.	B
4.	E	9.	C	14.	C	19.	E
5.	B	10.	E	15.	B	20.	D

Section 2

1. (a) 2 marks for symbols:

   • All correct   2
   • At least two different symbols correct   1

   1 mark for correct representation of external circuit wiring with no gaps

   (b) $V = IR$ 1
        $2.5 = 0.5 \times R$ 1
        $R = 5\ \Omega$

   (c) Mark for effect can only be awarded if a justification is attempted.

   Incorrect or no effect stated, regardless of justification – no marks.

   Effect:
   (It/lamp L is) brighter   1
   Justification:
   M is in parallel (with resistor)   1
   Greater current in/through lamp L (than that in M)   1

   OR

   Effect:
   (It/lamp L is) brighter   1
   Justification:
   M is in parallel (with resistor)   1
   Greater voltage across lamp L (than across M)   1

2. (a) (Graph) X   1
       An LED/diode/it only conducts in one direction   1

   (b) (i) $P = IV$ 1
            $P = 0.5 \times 4$
            $P = 2$(W)

            $E = Pt$ 1
            $E = 2 \times 60$ 1
            $E = 120$ J   1

       (ii) $Q = I \times t$ 1
            $Q = 0.5 \times 60$ 1
            $Q = 30\ C$ 1

3. (a) (i) 15 µs

       (ii) **Method 1:**
            $d = vt$ 1
            $= 5200 \times 15 \times 10^{-6}$ 1
            $= 0.078$ (m)   1

(If this line is the final answer then unit required for mark)

$$thickness = \frac{0 \cdot 078}{2}$$

$$= 0 \cdot 039 \text{ m} \qquad 1$$

Method 2:

$$time = \frac{15 \times 10^{-6}}{2}$$

$$= 7 \cdot 5 \times 10^{-6} \text{ (s)} \qquad 1$$

$$d = vt \qquad 1$$

$$= 5200 \times 7 \cdot 5 \times 10^{-6} \qquad 1$$

$$= 0 \cdot 039 \text{ m} \qquad 1$$

(b)

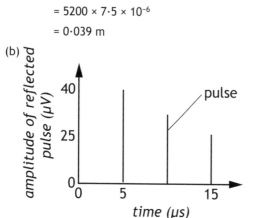

1 mark is awarded for "a peak at a time greater than 5 µs and less than 15 µs"

and

1 mark is awarded for "an amplitude greater than 25 µV and less than 40 µV"

(c) (i) This is a "show that" question so must start with correct formula or zero marks.

$$f = \frac{1}{T} \qquad 1$$

$$= \frac{1}{4 \cdot 0 \times 10^{-6}} \qquad 1$$

$$= 2 \cdot 5 \times 10^{5} \text{ Hz}$$

(ii) $v = f\lambda$ $\qquad 1$

$$5200 = 2 \cdot 5 \times 10^{5} \times \lambda \qquad 1$$

$$\lambda = 0 \cdot 021 \text{ m} \qquad 1$$

(d) Mark for effect can only be awarded if a justification is attempted.

Incorrect or no effect stated, regardless of justification — no marks.

(Speed of ultrasound in brass is) less (than in steel). $\qquad 1$

Takes greater time to travel (same) distance/thickness. $\qquad 1$

4.

Demonstrates no understanding	0 marks
Demonstrates limited understanding	1 mark
Demonstrates reasonable understanding	2 marks
Demonstrates good understanding	3 marks

This is an open-ended question.

1 mark: The student has demonstrated a limited understanding of the physics involved. The student has made some statement(s) which is/are relevant to the situation, showing that at least a little of the physics within the problem is understood.

2 marks: The student has demonstrated a reasonable understanding of the physics involved. The student

makes some statement(s) which is/are relevant to the situation, showing that the problem is understood.

3 marks: The maximum available mark would be awarded to a student who has demonstrated a good understanding of the physics involved. The student shows a good comprehension of the physics of the situation and has provided a logically correct answer to the question posed. This type of response might include a statement of the principles involved, a relationship or an equation, and the application of these to respond to the problem. This does not mean the answer has to be what might be termed an "excellent" answer or a "complete" one.

5. (a) Correctly labelled the angle of incidence **and** angle of refraction

(b) Decreases

(c) B

(d) $P = \dfrac{F}{A}$ $\qquad 1$

$$= \frac{61000}{1 \cdot 1 \times 10^{-5}} \qquad 1$$

$$= 5 \cdot 5 \times 10^{9} \text{ Pa} \qquad 1$$

6. (a) Increases

(b) (i) Mark for choice can only be awarded if an explanation is attempted.

Incorrect or no choice made, regardless of explanation — no marks.

Choice:

(source) X $\qquad 1$

Explanation:

beta (source required) $\qquad 1$

long half-life $\qquad 1$

(ii) Time for activity to (decrease by) half

OR

Time for half the nuclei to decay

(iii) (high frequency) electromagnetic wave

(c) 2 hours

7. (a) (i) **Using Pythagoras:**

$$\text{Resultant}^2 = (6 \cdot 0 \times 10^{3})^2$$

$$+ (8 \cdot 0 \times 10^{3})^2 \qquad 1$$

$$\text{Resultant} = 10 \times 10^{3} \text{ N} \qquad 1$$

Using scale diagram:

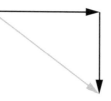

vectors to scale $\qquad 1$

Resultant = 10×10^{3} N $\qquad 1$

(allow $\pm 0 \cdot 5 \times 10^{3}$ N tolerance)

(ii) **Using trigonometry:**

$$\tan \theta = 6/8 \qquad 1$$

$$\theta = 37° \qquad 1$$

Using scale diagram:

angles correct   1

$\theta = 37°$ 1

(allow ±2° tolerance)

(iii) $F = ma$ 1

$10 \times 10^3 = 5\cdot 0 \times 10^6 \times a$ 1

$a = 2\cdot 0 \times 10^{-3}$ ms^{-2} 1

(b) **Upward arrow:** buoyancy force/upthrust/force of
water on ship/flotation force   1

Downward arrow: weight/force of gravity   1

(These) forces are balanced   1

8. (a) (i) • length/width of card   1

• time taken for card to pass (through) the
light gate   1

• time taken (for trolley to travel from starting
position) to light gate   1

(ii) reaction time (can cause error with the stop
clock reading)

OR

card may not have passed straight through light
gate

OR

length/width of card not measured properly (e.g.
ruler not straight along card)

OR

other suitable reason

(b) $a = \dfrac{v - u}{t}$ 1

$= \dfrac{1\cdot 6 - 0}{2\cdot 5}$ 1

$= 0\cdot 64$ ms^{-2} 1

9. (a) suitable curved path   1

(b) (i) $a = \dfrac{v - u}{t}$ 1

$9\cdot 8 = \dfrac{v - 0}{0\cdot 80}$ 1

$v = 7\cdot 8$ ms^{-1} 1

(ii) $\bar{v} = 3\cdot 9$ ms^{-1} 1

$d = \bar{v} t$ 1

$= 3\cdot 9 \times 0\cdot 80$ 1

$= 3\cdot 1$ m   1

(c) (it will take the) same (time)

10. Demonstrates no understanding   0 marks

Demonstrates limited understanding   1 mark

Demonstrates reasonable understanding   2 marks

Demonstrates good understanding   3 marks

This is an open-ended question.

1 mark: The student has demonstrated a limited
understanding of the physics involved. The student has
made some statement(s) which is/are relevant to the
situation, showing that at least a little of the physics
within the problem is understood.

2 marks: The student has demonstrated a reasonable
understanding of the physics involved. The student
makes some statement(s) which is/are relevant to the
situation, showing that the problem is understood.

3 marks: The maximum available mark would be
awarded to a student who has demonstrated a good
understanding of the physics involved. The student shows
a good comprehension of the physics of the situation and
has provided a logically correct answer to the question
posed. This type of response might include a statement
of the principles involved, a relationship or an equation,
and the application of these to respond to the problem.
This does not mean the answer has to be what might be
termed an "excellent" answer or a "complete" one.

11. (a) (i) $E_p = mgh$ 1

$E_p = 0\cdot 040 \times 9\cdot 8 \times 0\cdot 50$ 1

$E_p = 0\cdot 20$ J   1

(ii) kinetic (energy) to heat (and sound)

OR

kinetic (energy) of the marble to kinetic (energy)
of the sand.

(b) (i) suitable scales, labels and units   1

all points plotted accurately to ± half a division 1

best fit <u>curve</u>   1

(ii) Consistent with best fit curve from (b)(i).

(iii) Any two from:

• Repeat (and average)

• Take (more) readings in the 0·15 (m) to
0·35 (m) drop height range

• Increase the height range

• level sand between drops

• or other suitable improvement

(1) each

(c) (i) suitable variable

e.g.

• mass/weight of marble

• angle of impact

• type of sand

• diameter of marble

• radius of marble

• density of marble

• volume of marble

• speed of marble

• time of drop

(ii) How independent variable can be measured/
changed   1

State at least one other variable to be controlled 1

NATIONAL 5 PHYSICS 2016

Section 1

1.	C	6.	C	11.	A	16.	D
2.	A	7.	A	12.	B	17.	E
3.	D	8.	A	13.	B	18.	C
4.	C	9.	E	14.	E	19.	D
5.	B	10.	C	15.	D	20.	D

Section 2

1. (a) $Q = It$ 1
    $24 = I \times 0\cdot0012$ 1
    $I = 20\,000$ A   1

    (b) $24 \div 1\cdot6 \times 10^{-19}$
    $= 1\cdot5 \times 10^{20}$ (electrons)   1

    (c) (Metal strip) is a conductor   1

    (More) current/charge/electrons will pass through (the strip than building)   1

    NB it is not acceptable to say "lightning/electricity will pass through".

2. (a) Voltmeter across resistor R   1

    (b) Increase/decrease/vary/change the <u>resistance</u> of the <u>variable resistor</u>   1

    (c) **Numerical method:**
    Ohm's Law stated   1
    All substitutions shown   2

    5 Ω   1

    $V = IR$
    $1 = 0\cdot2 \times R$
    $R = 5\ (\Omega)$

    $V = IR$
    $2\cdot5 = 0\cdot5 \times R$
    $R = 5\ (\Omega)$

    $V = IR$
    $3\cdot2 = 0\cdot64 \times R$
    $R = 5\ (\Omega)$

    $V = IR$
    $6\cdot2 = 1\cdot24 \times R$
    $R = 5\ (\Omega)$

    (resistance of R = 5 Ω)

    NB it is not acceptable to total or average the voltages and currents. If this is done then 1 mark maximum for statement of Ohm's Law.

    Graphical method:
    Suitable scales and labels   1
    All points plotted accurately to ±half a division   2
    Line drawn and gradient calculated to be 5 Ω   1

    (d) (Resistance is) changing/not constant/increasing   1

3. (a) **NB answers to "Show that" questions must start with the correct equation or the answer will be awarded 0 marks.**
    $E_h = cm\Delta T$ 1
    $E_h = 4180 \times 6\cdot0 \times 25$ 1
    $E_h = 627\,000$ J

    (b) (i) $P = \dfrac{E}{t}$ 1

    $1800 = \dfrac{627\,000}{t}$ 1

    $t = 350$ s   1

(ii) <u>Heat</u> (energy) is lost (from the water) to the washing machine/drum/surroundings/clothing   1

OR

Some of the energy is used to <u>heat</u> up the washing machine/element/drum/clothing   1

(c) Voltage across thermistor decreases   1

MOSFET/transistor switches off/deactivates   1

Relay switches off/relay switch opens/relay deactivates   1

4. (a) (Black bulb) thermometer, photodiode, phototransistor, thermistor, thermocouple, CCD, thermochromic film   1

    (b) Gamma (radiation/rays)   1

    (c) (i) **NB answers to "Show that" questions must start with the correct equation or the answer will be awarded 0 marks.**

    $v = f\lambda$ 1
    $3\cdot0 \times 10^8 = 1\cdot2 \times 10^9 \times \lambda$ 1
    $\lambda = 0\cdot25$ m

    For alternative methods calculating v or f there must be final statement to show that calculated value of v or f is the same as the value stated in the question/data sheet to gain the second mark.

    (ii) Microwave (radiation)   1

5. **This is an open-ended question.**

Demonstrates no understanding	0 marks
Demonstrates limited understanding	1 mark
Demonstrates reasonable understanding	2 marks
Demonstrates good understanding	3 marks

1 mark: The student has demonstrated a limited understanding of the physics involved. The student has made some statement(s) which is/are relevant to the situation, showing that at least a little of the physics within the problem is understood.

2 marks: The student has demonstrated a reasonable understanding of the physics involved. The student makes some statement(s) which is/are relevant to the situation, showing that the problem is understood.

3 marks: The maximum available mark would be awarded to a student who has demonstrated a good understanding of the physics involved. The student shows a good comprehension of the physics of the situation and has provided a logically correct answer to the question posed. This type of response might include a statement of the principles involved, a relationship or an equation, and the application of these to respond to the problem. This does not mean the answer has to be what might be termed an "excellent" answer or a "complete" one.

6. (a) (i) Normal drawn <u>and labelled</u>   1

    (ii) Both angles indicated and labelled   1

    (b) (i) 8°   1

    (ii) Any single value between 40° and 42° inclusive   1

    (c) **Any one of:**
    To obtain more reliable results
    Eliminate rogue results/outliers
    To allow an average/mean to be calculated
    More accurate   1

7. (a)
$$A = \frac{N}{t}$$ 1

$$A = \frac{7 \cdot 92 \times 10^{18}}{900}$$ 1

$$A = 8 \cdot 8 \times 10^{15} \text{ Bq}$$ 1

(b) $8 \cdot 8 \times 10^{15} \times 4 \cdot 49 \times 10^{-14}$ 1

$= 400 \text{ W}$ 1

(c) **Any one of:**
(Alpha is) more easily absorbed/stopped/blocked
(Alpha) is absorbed by thinner materials/less dense materials
Gamma is absorbed by thicker materials/more dense materials
(Alpha) is less penetrating (than gamma)
Gamma is more penetrating (than alpha)   1

8. (a) (i) $D = \frac{E}{m}$ 1

$$D = \frac{9 \cdot 6 \times 10^{-5}}{0 \cdot 5}$$ 1

$$D = 1 \cdot 9 \times 10^{-4} \text{ Gy}$$ 1

(ii) $H = Dw_R$ 1

$H = 1 \cdot 9 \times 10^{-4} \times 1$ 1

$H = 1 \cdot 9 \times 10^{-4} \text{ Sv}$ 1

(b) No. of half-lives $= \frac{144}{36} = 4$ 1

$12 \rightarrow 6 \rightarrow 3 \rightarrow 1 \cdot 5 \rightarrow 0 \cdot 75$

Mark for evidence of activity halving   1

Final answer:
0·75 kBq   1

9. (a) (i) **Using Pythagoras:**
Resultant$^2 = 40^2 + 75^2$ 1
Resultant $= 85 \text{ m}$ 1

Using scale diagram:

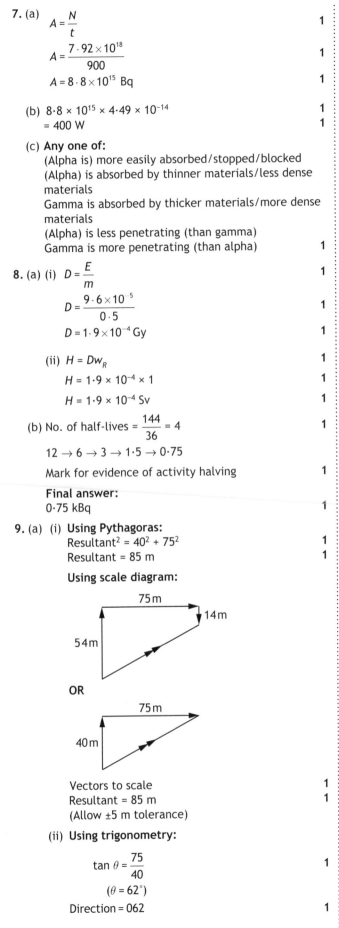

OR

Vectors to scale   1
Resultant $= 85 \text{ m}$ 1
(Allow ±5 m tolerance)

(ii) **Using trigonometry:**

$$\tan \theta = \frac{75}{40}$$ 1

$(\theta = 62°)$

Direction $= 062$ 1

Using scale diagram:

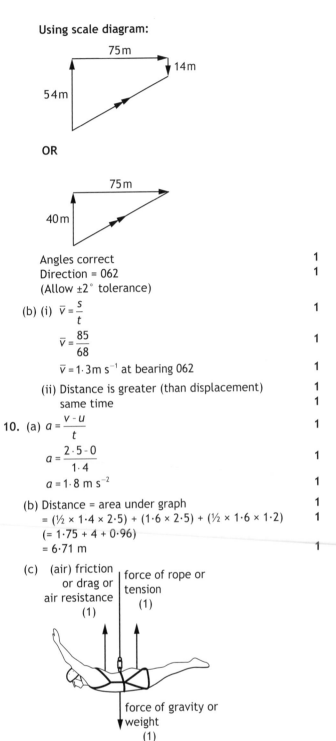

OR

Angles correct   1
Direction $= 062$ 1
(Allow ±2° tolerance)

(b) (i) $\bar{v} = \frac{s}{t}$ 1

$$\bar{v} = \frac{85}{68}$$ 1

$\bar{v} = 1 \cdot 3 \text{ m s}^{-1}$ at bearing 062   1

(ii) Distance is greater (than displacement)   1
same time   1

10. (a) $a = \frac{v - u}{t}$ 1

$$a = \frac{2 \cdot 5 - 0}{1 \cdot 4}$$ 1

$a = 1 \cdot 8 \text{ m s}^{-2}$ 1

(b) Distance = area under graph   1
$= (½ \times 1 \cdot 4 \times 2 \cdot 5) + (1 \cdot 6 \times 2 \cdot 5) + (½ \times 1 \cdot 6 \times 1 \cdot 2)$ 1
$(= 1 \cdot 75 + 4 + 0 \cdot 96)$
$= 6 \cdot 71 \text{ m}$ 1

(c) (air) friction or drag or air resistance (1) — force of rope or tension (1)

force of gravity or weight (1)

1 mark awarded for each force, correctly labelled and with corresponding direction.

11. This is an open-ended question.
Demonstrates no understanding   0 marks
Demonstrates limited understanding   1 mark
Demonstrates reasonable understanding   2 marks
Demonstrates good understanding   3 marks

1 mark: The student has demonstrated a limited understanding of the physics involved. The student has made some statement(s) which is/are relevant to the situation, showing that at least a little of the physics within the problem is understood.

2 marks: The student has demonstrated a reasonable understanding of the physics involved. The student makes some statement(s) which is/are relevant to the situation, showing that the problem is understood.

3 marks: The maximum available mark would be awarded to a student who has demonstrated a good understanding of the physics involved. The student shows a good comprehension of the physics of the situation and has provided a logically correct answer to the question posed. This type of response might include a statement of the principles involved, a relationship or an equation, and the application of these to respond to the problem. This does not mean the answer has to be what might be termed an "excellent" answer or a "complete" one.

12. (a) $W = mg$ 1

  $W = 3 \cdot 00 \times 10^3 \times 9 \cdot 8$ 1

  $W = 2 \cdot 9 \times 10^4\,\mathrm{N}$ 1

 (b) (i) Light (energy) → electrical (energy) 1

  (ii) Maximise the light received (from the Sun) (or similar) 1

  (iii) $E = Pt$ 1

  $E = 395 \times 2 \times 60 \times 60$ 1

  $E = 2 \cdot 8 \times 10^6\ \mathrm{J}$ 1

 (c) (i) (4 × 10 =) 40 N 1

  (ii) $m = 3 \cdot 00 \times 10^3 - 1 \cdot 00 \times 10^3$ 1

   $= 2 \cdot 00 \times 10^3\ (\mathrm{kg})$

   $a = \dfrac{F}{m}$ 1

   $a = \dfrac{40}{2 \cdot 00 \times 10^3}$ 1

   $a = 0 \cdot 02\ \mathrm{m\ s^{-2}}$ 1

13. (a) (Two) <u>nuclei</u> combine (to form a larger nucleus) 1

 (b) 5505 (°C) 1

 (c) **NB answers to "Show that" questions must start with the correct equation or, in this instance, the answer will be awarded 1 mark maximum for the speed of light.**

  $d = vt$ 1

  $d = 3 \cdot 0 \times 10^8 \times (365 \cdot 25 \times 24 \times 60 \times 60 \times 640)$ 1 + 1

  NB 1 for speed of light; 1 for all the numbers in the bracket.

  $d = 6 \cdot 1 \times 10^{18}\ \mathrm{m}$

 (d) The light/radiation from the explosion has not reached the Earth yet.

  OR

  The light/radiation takes time/640 years to reach Earth/to get here.

Acknowledgements

Permission has been sought from all relevant copyright holders and Hodder Gibson is grateful for the use of the following:

Image © Stuart Elflett/Shutterstock.com (2014 Section 2 page 6);
Image © Ints Vikmanis/Shutterstock.com (2014 Section 2 page 21);
Image © Sandra R. Barba/Shutterstock.com (2014 Section 2 page 24);
Image © Rob Byron/Shutterstock.com (2015 Section 1 page 10);
Image © MarcelClemens/Shutterstock.com (2015 Section 2 page 23);
Image © Procy/Shutterstock.com (2015 Section 2 page 25);
Image © Piotr Krzeslak/Shutterstock.com (2016 Section 2 page 6);
Image © AstroStar/Shutterstock (2016 Section 2 page 30).